Studies in Systems, Decision and Control

Volume 436

Series Editor

Janusz Kacprzyk, Systems Research Institute, Polish Academy of Sciences, Warsaw, Poland

The series "Studies in Systems, Decision and Control" (SSDC) covers both new developments and advances, as well as the state of the art, in the various areas of broadly perceived systems, decision making and control–quickly, up to date and with a high quality. The intent is to cover the theory, applications, and perspectives on the state of the art and future developments relevant to systems, decision making, control, complex processes and related areas, as embedded in the fields of engineering, computer science, physics, economics, social and life sciences, as well as the paradigms and methodologies behind them. The series contains monographs, textbooks, lecture notes and edited volumes in systems, decision making and control spanning the areas of Cyber-Physical Systems, Autonomous Systems, Sensor Networks, Control Systems, Energy Systems, Automotive Systems, Biological Systems, Vehicular Networking and Connected Vehicles, Aerospace Systems, Automation, Manufacturing, Smart Grids, Nonlinear Systems, Power Systems, Robotics, Social Systems, Economic Systems and other. Of particular value to both the contributors and the readership are the short publication timeframe and the worldwide distribution and exposure which enable both a wide and rapid dissemination of research output.

Indexed by SCOPUS, DBLP, WTI Frankfurt eG, zbMATH, SCImago.

All books published in the series are submitted for consideration in Web of Science.

George Chryssolouris · Kosmas Alexopoulos ·
Zoi Arkouli

A Perspective on Artificial Intelligence in Manufacturing

 Springer

George Chryssolouris ⓘ
Laboratory for Manufacturing Systems
and Automation
University of Patras
Rion Patras, Greece

Kosmas Alexopoulos ⓘ
Laboratory for Manufacturing Systems
and Automation
University of Patras
Rion Patras, Greece

Zoi Arkouli ⓘ
Laboratory for Manufacturing Systems
and Automation
University of Patras
Rion Patras, Greece

ISSN 2198-4182   ISSN 2198-4190 (electronic)
Studies in Systems, Decision and Control
ISBN 978-3-031-21830-9   ISBN 978-3-031-21828-6 (eBook)
https://doi.org/10.1007/978-3-031-21828-6

This Springer imprint is published by the registered company Springer Nature Switzerland AG
The registered company address is: Gewerbestrasse 11, 6330 Cham, Switzerland

Preface

Undoubtedly, Manufacturing is of great importance throughout a number of industrial sectors. Manufacturing is an activity that creates jobs and wealth and is widely regarded as a key pillar of both social and economic development.

From a technology perspective, manufacturing is both a testbed for new technologies, whilst simultaneously, it is the cause of major technological developments, typically geared towards addressing manufacturing challenges.

Along these lines, manufacturing environments can be testbeds for Artificial Intelligence (AI) concepts, technologies and tools, while they can also help for new developments in AI that address problems in the manufacturing arena.

This book offers an AI perspective, from a manufacturing point of view by providing a general vision as well as applications and AI solutions to a number of manufacturing challenges. It presents what is emerging in the AI field, related to manufacturing and what the future could possibly look like.

Rion Patras, Greece
December 2022

George Chryssolouris
Kosmas Alexopoulos
Zoi Arkouli

Acknowledgement

This book summarizes the work that is being done, on Artificial Intelligence (AI) and Manufacturing in recent years, at our Laboratory for Manufacturing Systems and Automation (LMS), University of Patras. This work is the continuation of that carried out at MIT, in the 80s, by one of the authors (George Chryssolouris). The book attempts to provide a perspective of Artificial Intelligence from a Manufacturing point of view. It introduces three levels of interaction with AI technology, which may help in understanding, applying, and also developing AI technology, related to manufacturing.

There are a number of projects, related to the work described in this book, funded by the European Commission and we would like to acknowledge and thank the help of the European Commission (EC) and, in particular, the EC research project "Multi-Agent Systems for Pervasive Artificial Intelligence for assisting Humans in Modular Production Environments"—"**MAS4AI**" (Grant Agreement: 957204). Finally, the authors would like to thank Prof. Panos Stavropoulos and Dr. Sotiris Makris for their substantial contribution to the preparation of this book.

George Chryssolouris
Kosmas Alexopoulos
Zoi Arkouli

Contents

Chapter 1
Introduction

Abstract This chapter introduces Artificial Intelligence (AI) as well as manufacturing and draws the links between them. AI technology has been used in manufacturing for decades, on complex decision making, in support of the manufacturers, in their business processes. AI has been studied since the 1940s, however, it is only recently that scientists and industry practitioners are getting closer to effectively exploiting its potential since AI technologies have become more mature and affordable. In principle, AI can be understood as *"software (and possibly also hardware) systems, designed by humans that, given a complex goal, act in the physical or digital dimension by perceiving their environment through data acquisition, interpreting the collected structured or unstructured data reasoning on the knowledge, or processing the information, having derived from this data and deciding on the best action(s) to be taken to achieve the given goal"*. The historical evolution of AI is presented in four upsurge periods, driven by improvements in relevant technologies. At the same time, manufacturing can be defined as the transformation of materials and information into goods for the fulfilment of human needs and it is one of the primary wealth creation activities for any nation besides contributing significantly to employment. Decision -making in manufacturing considers four classes of attributes, namely cost, time, quality, and flexibility. These attributes are discussed and examples of AI applications, related to optimization and decision making of these attributes, are provided. At the end a three-layered taxonomy into process, equipment and systems are presented. This taxonomy will be used for discussing applications of AI in the next chapters.

Keywords Decision-making · Decision-making criteria · Manufacturing

1.1 Introduction

Artificial Intelligence (AI) has been studied since the 1940s, but it seems that only recently are scientists getting closer to effectively exploiting its potential. In particular, within the 80 years of AI research, there have been periods of upsurge and downturn, according to the restrictions of the contemporary technology e.g., difficulty in handling non-linear problems, the necessity for human expertise during

G. Chryssolouris et al., *A Perspective on Artificial Intelligence in Manufacturing,*
Studies in Systems, Decision and Control 436,
https://doi.org/10.1007/978-3-031-21828-6_1

feature extraction, etc. Since AI has become a critical aspect of modern societies, it is important that a widespread understanding of key concepts, technologies, and applications be developed.

Artificial Intelligence can be generally understood as *"the intelligence exhibited by machines or software"* as defined in Wikipedia. Albus (1991) defined intelligence as *"the ability of a system to act appropriately in an uncertain environment, where the appropriate action is that which increases the probability of success, whilst success is the achievement of behavioral sub-goals that support the system's ultimate goal."* [1]. In 1956, John McCarthy defined AI as *"the science and engineering of making intelligent machines"*. A stricter definition was given by Russel and Norvig, who defined AI as *"the study and design of intelligent agents, where an intelligent agent is a system that perceives its environment and takes actions that maximize its chances of success"* [2]. Moreover, there are four schools of thought according to Russel and Norvig: systems that think humanly (cognitive approach), systems that think rationally (laws of thoughts), systems that act humanly (imitation game), and systems that act rationally (intelligent agents). *"Artificial intelligence (AI) refers to systems that display intelligent behaviour by analysing their environment and taking actions – with some degree of autonomy – to achieve specific goals. AI-based systems can be purely software-based, acting in the virtual world (e.g. voice assistants, image analysis software, search engines, speech and face recognition systems) or AI can be embedded in hardware devices (e.g. advanced robots, autonomous cars, drones, or Internet of Things applications)"* [3]. The EC High-Level Expert Group on Artificial Intelligence has provided an illustration of an AI system, as depicted in Fig. 1.1 [4], whereas, the latest definition of AI on behalf of the European Commission has been published by the same group and it is as follows: *"Artificial intelligence (AI) systems are software (and possibly also hardware) systems designed by humans that, given a complex goal, act in the physical or digital dimension by perceiving their environment through data acquisition, interpreting the collected structured or unstructured data, reasoning on the knowledge, or processing the information, derived from this data and deciding on the best action(s) to be taken to achieve the given goal. AI systems can either use symbolic rules or learn a numeric model, and they can also adapt their behaviour by analysing the ways that the environment is affected by their previous actions. As a scientific discipline, AI includes several approaches and techniques, such as machine learning (of which deep learning and reinforcement learning are specific examples), machine reasoning (which includes planning, scheduling, knowledge representation and reasoning, search, and optimization), and robotics (which includes control, perception, sensors and actuators, as well as the integration of all other techniques into cyber-physical systems)"* [4].

A classification of AI into seven distinct types, based on the degree that an AI system can replicate human capabilities, and in more detail, based on their likeness to the human mind has been proposed by Joshi (2019): reactive machines (they are capable of responding to different kinds of stimuli' they do not have memory i.e. they cannot use previously gained experience to inform present actions, they are used to automatically respond to a limited set of inputs), limited memory machines (in addition to the capabilities of the reactive machines, they are capable of learning

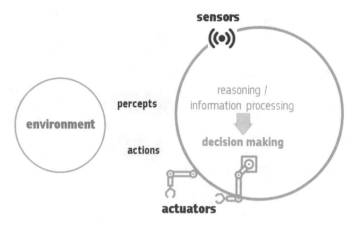

Fig. 1.1 An AI system depiction as provided by AI HLEG [4]

from historical data to make decisions), theory of mind systems (they are currently under development by researchers, exist in concept or work in progress, they should be able to better understand the entities they are interacting with, by discerning needs, emotions, beliefs, and thought processes), self-aware (it currently exists only hypothetically, since an AI that has evolved to be so akin to the human brain has developed self-awareness) [5]. AI can also be classified into Artificial Narrow Intelligence, Artificial General Intelligence, and Artificial Superintelligence [6]. Artificial Narrow Intelligence represents all the existing AI, it refers to systems that can perform tasks autonomously and their actions are restricted in what they have been programmed to do (reactive and limited memory machines) and Artificial General Intelligence which is the ability of an AI agent to perceive, understand, learn and function exactly like a human. In this case, multiple competencies are independently built and form connections and generalizations across domains, which in turn, massively cuts down on the necessary training time [7]. On the other hand, artificial superintelligence will not only replicate the multi-faceted intelligence of humans, but it will perform better at everything it is being done thanks to its capabilities for overwhelmingly greater memory, faster data processing and analysis, as well as decision-making [8].

The AI evolution comprises four stages: the infancy period (the 1940s), the upsurge period (1960s), the second upsurge period (1980s), and the third boom period (after the 2000s). The infancy period started approximately in 1940 when the McCulloch-Pitts Neuron model and the Herb rule were proposed for the discussion of the way that neurons work in the human brain. Developments of capabilities such as playing chess games and solving simple problems that had started.

The first upsurge period started around 1960. The major contributions to this period's AI technology were those of the Perceptron model, proposed for the simulation of the nervous system of human learning, with linear optimization (1959), and the Adaptive Linear unit, which is a network model used in practical applications (communication, weather forecasting). The limitations which dominated the early AI were due to the difficulty in handling non-linear problems.

The solution to this difficulty came with the second upsurge period, which started around 1980 namely, the development of the Hopfield network circuit, along with the Back Propagation algorithm, the Boltzman Machine, the Support Vector Machine, the Restricted Boltzman Machine and the Auto Encoder. After the non-linearity had been faced, the need for human expertise, during feature extraction and the dependence of the performance on the engineering features needed also to be handled.

The present third boom period, which started after 2000, deals with the limitations of the previous period mostly by taking advantage of deep learning models. Some worth mentioning innovations of the third boom are Recurrent Neural Networks, Long Short-term Memory (LSTM), Convolutional Neural Networks (CNNs), Deep Belief Networks, Deep Auto Encoder, Deep Boltzman Machine, Denoising Auto Encoder, Deep Convolutional Neural Networks, and Attention-based LSTM. Furthermore, research interest in Machine Learning (ML) has intensified. ML is deemed one of the long-term goals of the AI community and refers to the gathering of new knowledge, as well as the development of motor and cognitive skills through instructions and practice [9]. ML can be branched into three main categories: supervised learning, unsupervised learning, and reinforcement learning, depending on the conditions required for the learning process to be held. These categories can serve as a high-level taxonomy, but they are not necessarily mutually exclusive. For instance, hybrid approaches such as semi-supervised learning also exist and can belong to more than one branch of this taxonomy. ML approaches are able to pinpoint highly complex and non-linear patterns in raw data of several types and sources, creating models for activities such as classification, detection, or prediction [10].

The advances in information technologies (communication, big-data management, and computing power) are the key enablers of the boost in AI developments, which have already been integrated into many applications of everyday life. For instance, personal assistants e.g. Siri [11], Alexa [12], suggestive searches, such as Google's autocompleting [13], and automatically steering, accelerating, and braking vehicles e.g. Tesla's Autopilot [14]. Moreover, real-time conversational guidance, such as Cogito [15], as well as recommendation systems for amusement, including Pandora [16] for music, Netflix for movies, and Amazon for purchases, as well as smart household devices, such as Nest thermostat [17] are only some of the most popular AI developments, which are already in use. Besides the everyday life of individuals, it is worth exploring the areas where AI can contribute, the advances compared to the already existing approaches of the manufacturing practice, as well as the existing integration of AI tools into manufacturing systems. In this scope, the next paragraphs are dedicated to presenting fundamental definitions, related to manufacturing, as well as an introduction to decision-making in manufacturing.

"Manufacturing is defined as the transformation of materials and information into goods for the fulfilment of human needs and it is one of the primary wealth creation activities for any nation and contributes significantly to employment" [18]. By way of illustration, Fig. 1.2 depicts the added value of manufacturing per country as a percentage of the gross domestic product (GDP), where for the majority of the studied countries the value added exceeds 9%.

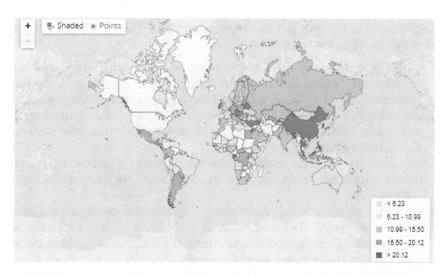

Fig. 1.2 Manufacturing, value added (% GDP) 2021 [19], Reprinted under the CC BY-4.0 license terms

As a result of the definition of manufacturing, it can be claimed that manufacturing is a system whose inputs are the customers' needs, together with the designers' creativity that led to the starting point of production, i.e. product design (Fig. 1.3). The system's output is the delivery of finished products to the market [20]. Decision-making is often required in the context of manufacturing systems.

In the context of manufacturing, a decision is basically the selection of values for certain decision variables either related to the design or to the operation of a manufacturing process, machine, or system. This decision-making in turn requires technical understanding and expertise, as well as the ability to satisfy particular business objectives. Hence, manufacturing is deemed as a multi-disciplinary field requiring engineering but also management skills and competencies. In this view, it is useful to define a taxonomy aiming to facilitate the identification of issues and to

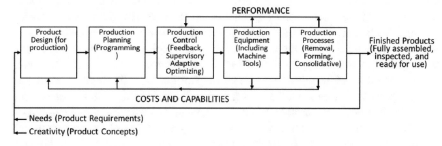

Fig. 1.3 Manufacturing systems [18], Reprinted with permission from Springer-Verlag New York INC

allow a scientific approach to the encountered problems. The following subdivision of manufacturing is suggested:

- *Design and manufacturing interfaces*: design and manufacturing interfaces include user interfaces that are utilized during design, planning, programming, production monitoring, inspection, etc. These interfaces, serve among others, the communication of the features and characteristics of the products to be manufactured that in current practice is usually performed via CAD files, virtual or even physical prototypes.
- *Manufacturing processes*: manufacturing processes refer to the set of processes that are responsible for altering the form, shape and/or physical properties of a given material.
- *Manufacturing equipment*: manufacturing equipment refers to the pieces of hardware that are practically the embodiment of the manufacturing processes and perform the required physical actions to achieve the altering of the form, shape, and/or physical properties of the given materials.
- *Manufacturing systems*: manufacturing systems can be defined as the combination of manufacturing equipment and humans, bound by a common material and information flow. It is noted that the design of a system is interwoven with its operation.
- *Production planning*: production planning refers to the aggregate timing of production, and it dictates the flow of materials and information into the manufacturing system, as well as the workload of the production system. In short, a successful production plan is expected to lead to the production of sufficient amounts of products on time.
- *Production control*: production control is responsible for ensuring the proper execution of the manufacturing production plan. It collects data from the shop floor and directs a flow of information back to the planning level, closing the loop between the planning and execution phases. In more detail, the production control deals with the resources' coordination, but also with re-organization/re-allocation in case of disruptions e.g. machine breakdown, or demand variations.

When making decisions in manufacturing, four classes of attributes should be considered: namely, cost, time, quality, and flexibility. The set of these attributes is referred to as the manufacturing Tetrahedron to emphasize the interrelationship among them (Fig. 1.4). It is noted that the relevance and importance of these attributes might differ depending on the particular case and decision to be made. Decision-making in manufacturing systems is based on performance requirements, which specify the values of the relevant manufacturing attributes. In this light, the decision-making process can be regarded as a mapping from desired attribute values onto corresponding decision variable values. The scientific foundation of this mapping is usually based on techno-economical models, in which the levels of manufacturing attributes are related to the levels of different decision variables. As for the overall outcome of a manufacturing decision, it is rather a trade-off between the different manufacturing attributes, given that in most cases it is not possible to simultaneously optimize cost, time, quality, and flexibility. The AI decision-support systems could also be used for trade-off optimization. The assessment of these trade-offs imposes the evaluation of each attribute. It is noted that in current practice, the majority of quantitative measures in manufacturing, refer to cost- and time-related attributes, whereas more details about the manufacturing attributes are discussed in the next paragraphs.

Cost

Costs related to manufacturing encompass a number of different factors, which can be broadly classified into the following categories:

- *Equipment and facility costs* (e.g. costs of equipment for the operation of manu-facturing processes, the facilities used to house the equipment, the factory infrastructure, etc.).
- *Materials* (e.g. costs of raw materials, tools and auxiliary materials, such as coolants and lubricants to produce the product).
- *Labor* (costs deriving from the direct labor needed for the operation of the equipment and facilities).
- *Energy* (costs for the performance of the different processes).
- *Maintenance and training* (costs for labor, spare parts, etc. required to maintain the equipment, facilities, and systems, as well as the training costs to accommodate new equipment and technology).
- *Overhead* (indirect costs to support the manufacturing infrastructure).
- *Capital* (costs deriving from loans that are taken when it is not possible to cover the financial issues that occur with readily available capital within the manufacturing firm).

Time

Time attributes can be analyzed into the system's production rate i.e. how quickly a product can be produced, but also into the responsiveness of manufacturing to changes in design, volume demand, etc., which, however, will be discussed under the flexibility attribute. The production rate impacts, to some degree, all the other types of attributes. To be more specific, higher production rates typically result in

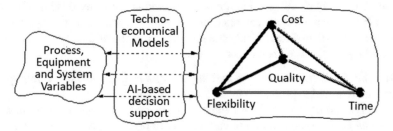

Fig. 1.4 Mapping Manufacturing Decision Variables into the Manufacturing Attributes [18], Reprinted with permission from Springer-Verlag New York INC

lower costs and potentially lower quality. Moreover, to achieve high production rates, it is often necessary to resort to automation, which may impact the system's flexibility.

Several notions are related to the time attribute and are used in industrial practice. For instance, the *theoretical production rate* of a system, or *machine cycle*, refers to the ideal scenario when a machine or system is running with no interruptions or delays and is the pieces produced per unit of time. The events that may arise and constrain this rate are typically related to the physics of the processes and the robustness of the machines. On the other hand, the actual production rate of a system is called *the process or system yield*, which is the number of acceptable pieces produced per unit of time, which incorporates delays and unpredicted interruptions e.g. machine breakdowns. The ratio of process yield or system yield as a percent of the machine cycle reflects the efficiency of the overall process, regarding the production rate.

The identification of achievable production rates in a manufacturing system requires the definition of several terms that have to do with the reliability of the equipment and the overall structure of the system. Reliability can be defined as the probability of a system or a component to perform its required function. The *failure rate* of a component is the ratio of the number of failures over a particular time period, whereas, the mean time between failure (MTBF) is the reciprocal of the failure rate. Availability is often a more informative indicator of breakdown behavior than reliability is, because it incorporates mean time to repair (MTTR). The availability of a system with a long MTBF can be deemed insensitive to the duration of the MTTR. On the contrary, the availability of a system with a short MTBF is very sensitive to the duration of the MTTR.

Flexibility

Flexibility, as already mentioned, refers to the responsiveness of a manufacturing system to changes in the target throughput, availability of resources, etc. Given the increasingly diversified customer base, and the decreasing lot sizes that govern demand nowadays, flexibility has stood out as one of the most critical attributes. Flexibility, however, cannot be properly considered in the decision-making process unless it is defined quantitatively. Although there is research academic work on the quantification of flexibility, industrial applications have been meagre [21]. Academic research has focused on one-of-a-kind or small lot size production systems, e.g. aerospace industry paradigms, but not on mass production systems. This is because the flexibility debate has concentrated on the ability of a manufacturing system to produce a range of products quickly and economically. The different aspects or types of flexibility [22] is listed hereunder:

- *Machine flexibility*: the ease of making changes in the machine's configuration that are required to produce a given set of part types.
- *Process flexibility*: the ability to produce a given set of part types in different ways.
- *Product flexibility*: the ability to change over to produce new products economically and quickly.
- *Routing flexibility*: the ability to handle breakdowns and to continue producing a given set of part types.

- *Volume flexibility*: the ability to operate profitably at different production volumes.
- *Expansion flexibility*: the ability to expand the system easily and in a modular fashion.
- *Operation flexibility*: the ability to interchange the ordering of several operations for each part type.
- *Production flexibility*: the set of part types that the manufacturing system can produce.

A different approach to viewing flexibility is to link it to resilience, where resilience refers to the ability to withstand disruptions, without the incurrence of significant additional costs. An approach, based on the penalty of change measurement method, has been proposed for the quantification of resilience in a manufacturing system and the enabling of decision-makers to assess alternatives for strategic investments [23].

Quality

The quality attribute is also difficult to be quantitatively expressed as it is broadly related to customer satisfaction, which in turn, depends on multiple factors, including the features of a product, its maintainability, and a host of other subjective factors. Nevertheless, customer satisfaction can be traced to two major factors at the origin of a product: its design and manufacture.

In manufacturing, quality typically refers to the degree that production meets design specifications, and is an aggregate of the quality of individual features and properties, i.e. geometric characteristics and the physical and/or chemical properties of the materials making up the product. Quality is typically measured from the most aggregate level, in terms of acceptance or rejection of a product, down to the elementary characteristics of a component. More elementary characteristics, such as dimensions and physical properties e.g. hardness and strength, are easier to be quantified, and thus provide an easier measuring task for testing and inspection, which is usually employed during the production process or immediately thereafter.

AI in manufacturing

Artificial Intelligence has already been employed to support decision-making and optimization in manufacturing systems since the 1970s. Focusing on the applications of the AI methods that were presented in manufacturing, in the period from 1970 to 1990, researchers focused on expert systems that were used to support the decision-making of people without expertise in a field and helped them decide as though they were experts [24]. Also, AI was frequently used for Manufacturing System Simulators in order to assign production tasks to the system resources, according to certain dispatching rules [25]. AI and the operations research approach to modeling flexible manufacturing systems, have been investigated in several use case scenarios, including machine layout and task scheduling [26].

Although AI was used in industry, there were no industrial booms, so the interest in AI started decreasing. This changed after 1990 thanks to the Artificial Neural Networks (ANN), the emergence of intelligent agents, and the availability of very large datasets, but mostly thanks to deep learning, which put AI in the spotlight again. Chryssolouris et al. [27] reviewed the contemporary decision-making in manufacturing, which remains one of the most interesting areas of AI in manufacturing, as indicated several years later by Sharp et al. [28]. In particular, in the latter review, an increasing interest in ML in manufacturing, especially for knowledge management, decision support, and lifecycle management is reported, while it is recognized that the availability of data makes ML a promising tool for more lean, agile, and energy-efficient manufacturing systems.

The next generation of industry-focused research on the development and implementation of more intelligent manufacturing systems. In this context, programs such as "smart manufacturing" (United States) and "Industrie 4.0" (Germany) have been established in several countries with the aim to reinforce domestic industry over the worldwide competition. At the same time, efforts have been made in defining architectures that will enable the necessary information flows by linking the operational technology and information technology domains, where heterogenous data from several sources e.g. machines, factory automation, supply chain management, etc. while enabling end-to-end communication among all production-relevant assets and fulfilling time and batch size constraints. Additional challenges include interoperability [29], compatibility with legacy industrial systems [30], as well as security, trust, and privacy that remain open research topics [31–33]. Reference architecture Model for Industry 4.0 (RAMI4.0), Industrial Internet Reference Architecture (IIRA), Stuttgart IT Architecture for Manufacturing (SITAM), IBM Industry 4.0, and Lasim Smart Factory (LASFA) are some of the reference architectures that have been proposed to cover these needs. Figure 1.5 depicts the RAMI4.0 architecture, which has been defined for "Industrie 4.0" as an initial step for a consensual base of the upcoming developments [34].

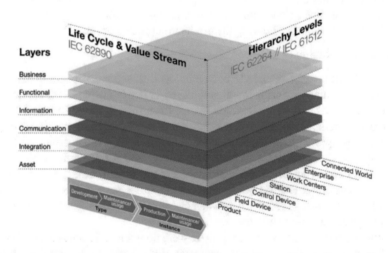

Fig. 1.5 Reference architecture for "Industrie 4.0", RAMI4.0 [34], Reprinted with permission from Plattform Industrie 4.0

As artificial intelligence technologies have become more mature and affordable, new applications have been introduced into production systems, in support of the manufacturers, on complex decision making and in their business processes. For instance, the potential of ML have been thoroughly investigated for quality inspection. ANNs are among the most popular classifiers and they have been employed for many applications such as fault diagnosis [35–37]. Moreover, Sharp et al. (2018) used Natural Language Processing (NLP) to investigate the effort being put towards advancing Machine Learning in manufacturing, as well as the prominent areas of ML use, the ML popular algorithms and have highlighted the existing gaps and areas, where ML could play a vital role [28]. Wuest et al. (2016) in their review on ML in manufacturing, have highlighted challenges, such as the manufacturing systems' dynamic nature, chaotic structures and high dimensionality, apart from foreseeing the upsurge of unsupervised methods with the increase of the available data [10].

On the other hand, Wang et al. (2018) have focused on deep learning for smart manufacturing, they have examined the evolution of deep learning technologies and their advantages over traditional ML, and pinpointed some limitations, which could be resolved with the evolution of computing resources [38]. In deep learning, the features are learnt with the transformation of data into abstract representations. In addition, the deep learning models have an end-to-end high hierarchical model structure, with the nonlinear combination of multi-layers, which enables the joint training of the parameters. On the other hand, the construction of the traditional machine learning models imposes step-by-step training.

Today's factories envision levels of automation, where industrial robots mimic the movements and, seemingly, the intentionality of human workers. Robots nowadays, not only do they work faster and more reliably than their human counterparts do, but also perform tasks, such as the microscopically precise assemblies, which are beyond the human capability altogether. The real benefit from AI in manufacturing, will not just be through the automation of tasks but through the provision of new levels of autonomy that will make entirely new applications possible and introduce new business processes in manufacturing. Predictive maintenance [39–41], energy management optimization [42, 43], involvement of customers in the design phase [44–46], plant-wide control [47] are some of the AI-enhanced capabilities. Moreover, automated processes, autonomous vehicles, robotics, Human–Robot collaboration and the combination of all the aforementioned may contribute to achieving the optimum dynamic exploitation of the resources, if integrated into combination with smart planning and scheduling systems [48].

This book will present a perspective on Artificial Intelligence in Manufacturing, following the taxonomy of AI applications depicted in Fig. 1.6. This taxonomy has emanated from the consideration of the different hierarchical levels of manufacturing systems e.g. factory, job shop, work center, resource. Each one of these levels is characterized by different planning and decision horizon levels, which in turn, pose different requirements for the design, implementation, and performance evaluation of the corresponding AI applications. The higher the hierarchical level, the longer the decision horizon, and the greater the impact of the decision [25]. In more detail,

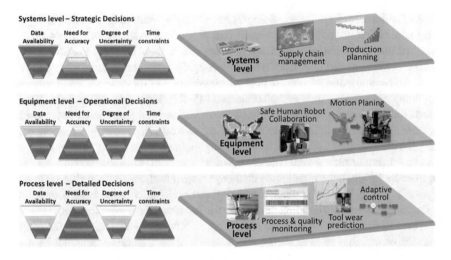

Fig. 1.6 AI-based applications in the systems-equipment-process levels taxonomy and decision-making characteristics per level

applications related to manufacturing processes (Chap. 2), manufacturing equipment (Chap. 3), as well as manufacturing systems (Chap. 4) are presented.

References

1. Albus, J.S.: Outline for a theory of intelligence. IEEE Trans. Syst. Man Cybern. **21**, 473–509 (1991). https://doi.org/10.1109/21.97471
2. Russel, S.J., Norvig, P.: Artificial Intelligence—A Modern Approach, 2nd edn (2003)
3. European Comission: Communication from the Commission: Artificial Intelligence for Europe. https://eur-lex.europa.eu/legal-content/EN/TXT/?uri=COM%3A2018%3A237%3AFIN
4. High-Level Expert Group on AI: A Definition of AI: Main Capabilities and Scientific Disciplines (2018)
5. Joshi, N.: 7 Types of Artificial Intelligence. https://www.forbes.com/sites/cognitiveworld/2019/06/19/7-types-of-artificial-intelligence/#100ef203233e
6. McLean, S., Read, G.J.M., Thompson, J., Baber, C., Stanton, N.A., Salmon, P.M.: The risks associated with Artificial General Intelligence: a systematic review. https://doi.org/10.1080/0952813X.2021.1964003 (2021)
7. Mahler, T.: Regulating Artificial General Intelligence (AGI), pp. 521–540 (2022). https://doi.org/10.1007/978-94-6265-523-2_26
8. Narain, K., Swami, A., Srivastava, A., Swami, S.: Evolution and control of artificial superintelligence (ASI): a management perspective. J. Adv. Manage. Res. **16**, 698–714 (2019). https://doi.org/10.1108/JAMR-01-2019-0006/FULL/PDF
9. Camastra, F., Vinciarelli, A.: Machine Learning for Audio, Image and Video Analysis: Theory and Applications. SPIE-Intl. Soc. Opt. Eng. (2015)
10. Wuest, T., Weimer, D., Irgens, C., Thoben, K.D.: Machine learning in manufacturing: advantages, challenges, and applications. Prod. Manuf. Res. **4**, 23–45 (2016). https://doi.org/10.1080/21693277.2016.1192517
11. Apple: Siri—Apple. https://www.apple.com/siri/

12. Amazon: Keyword Research, Competitive Analysis, & Website Ranking | Alexa. https://www.alexa.com/
13. Sullivan, D.: How Google Autocomplete Works in Search. https://www.blog.google/products/search/how-google-autocomplete-works-search/
14. Tesla Inc: Autopilot AI. https://www.tesla.com/autopilotAI
15. Cogito: Home Cogito. https://www.cogitocorp.com/
16. Pandora Media: Pandora—Music Genome Project ®. https://www.pandora.com/about/mgp
17. Google: Nest | Create a Connected Home. https://nest.com/
18. Chryssolouris, G.: Manufacturing Systems: Theory and Practice. Springer (2006)
19. World Bank National Accounts, OECD National Accounts: Manufacturing, Value Added (% of GDP) | Data. https://data.worldbank.org/indicator/NV.IND.MANF.ZS?end=2020&start=2020&type=shaded&view=map&year=2021
20. Merchant, E.M.: NSF Report on Research Priorities for the NSF Strategic Manufacturing Research Institute (1987)
21. Alexopoulos, K., Papakostas, N., Mourtzis, D., Chryssolouris, G.: A method for comparing flexibility performance for the lifecycle of manufacturing systems under capacity planning constraints, **49**, 3307–3317 (2010). https://doi.org/10.1080/00207543.2010.482566
22. Browne, J., Dubois, D., Rathmill, K., Sethi, S.P., Stecke, K.E.: Classification of flexible manufacturing systems. The FMS Mag. **2**, 114–117 (1984)
23. Alexopoulos, K., Anagiannis, I., Nikolakis, N., Chryssolouris, G.: A quantitative approach to resilience in manufacturing systems (2022). https://doi.org/10.1080/00207543.2021.2018519
24. Fox, J.: Expert systems and theories of knowledge. In: Boden, M.A. (ed.) Artificial Intelligence, pp. 157–181. Academic Press (1996)
25. Chryssolouris, G., Wright, K., Pierce, J., Cobb, W.: Manufacturing systems operation: dispatch rules versus intelligent control. Robot. Comput. Integr. Manuf. **4**, 531–544 (1988). https://doi.org/10.1016/0736-5845(88)90026-9
26. Heragu, S.S., Kusiak, A.: Machine layout problem in flexible manufacturing systems **36**, 258–268 (1988). https://doi.org/10.1287/OPRE.36.2.258
27. Chryssolouris, G., Graves, S., Ulrich, K.: Decision making in manufacturing systems: Product design, production planning, and process control. In: Proceedings of the 1991 NSF Design and Manufacturing systems conference (1991)
28. Sharp, M., Ak, R., Hedberg, T.: A survey of the advancing use and development of machine learning in smart manufacturing. J. Manuf. Syst. **48**, 170–179 (2018). https://doi.org/10.1016/j.jmsy.2018.02.004
29. Galati, F., Bigliardi, B.: Industry 4.0: emerging themes and future research avenues using a text mining approach. Comput. Ind. **109**, 100–113 (2019). https://doi.org/10.1016/J.COMPIND.2019.04.018
30. Givehchi, O., Landsdorf, K., Simoens, P., Colombo, A.W.: Interoperability for industrial cyber-physical systems: an approach for legacy systems. IEEE Trans. Industr. Inform. **13**, 3370–3378 (2017). https://doi.org/10.1109/TII.2017.2740434
31. Bicaku, A., Schmittner, C., Delsing, J., Maksuti, S., Palkovits-Rauter, S., Tauber, M., Matis-chek, R., Mantas, G., Thron, M.: Towards trustworthy end-to-end communication in industry 4.0 SECCRIT-secure cloud computing for critical infrastructure IT view project towards trustworthy end-to-end communication in industry 4.0 (2017)
32. Fraile, F., Tagawa, T., Poler, R., Ortiz, A.: Trustworthy industrial IoT gateways for interoperability platforms and ecosystems. IEEE Internet Things J. **5**, 4506–4514 (2018). https://doi.org/10.1109/JIOT.2018.2832041
33. Petroulakis, N.E., Lakka, E., Sakic, E., Kulkarni, V., Fysarakis, K., Somarakis, I., Serra, J., Sanabria-Russo, L., Pau, D., Falchetto, M., Presenza, D., Marktscheffel, T., Ramantas, K., Mekikis, P.V., Ciechomski, L., Waledzik, K.: SEMIoTICS architectural framework: end-to-end security, connectivity and interoperability for industrial IoT. In: Global IoT Summit, GIoTS 2019—Proceedings (2019). https://doi.org/10.1109/GIOTS.2019.8766399
34. VDI/VDE-Gesellschaft Mess- und Automatisierungstechnik: Status Report: Reference Architecture Model Industrie 4.0 (RAMI4.0) (2015)

35. Rodríguez, J.A., El Hamzaoui, Y., Hernández, J.A., García, J.C., Flores, J.E., Tejeda, A.L.: The use of artificial neural network (ANN) for modeling the useful life of the failure assessment in blades of steam turbines. Eng. Fail. Anal. **35**, 562–575 (2013). https://doi.org/10.1016/J.ENG FAILANAL.2013.05.002
36. Ahmadzadeh, F., Lundberg, J.: Remaining useful life prediction of grinding mill liners using an artificial neural network. Miner. Eng. **53**, 1–8 (2013). https://doi.org/10.1016/J.MINENG.2013.05.026
37. Ben Ali, J., Chebel-Morello, B., Saidi, L., Malinowski, S., Fnaiech, F.: Accurate bearing remaining useful life prediction based on Weibull distribution and artificial neural network. Mech. Syst. Signal Process. **56–57**, 150–172 (2015). https://doi.org/10.1016/J.YMSSP.2014.10.014
38. Wang, J., Ma, Y., Zhang, L., Gao, R.X., Wu, D.: Deep learning for smart manufacturing: methods and applications. J. Manuf. Syst. **48**, 144–156 (2018). https://doi.org/10.1016/j.jmsy.2018.01.003
39. Wu, S.J., Gebraeel, N., Lawley, M.A., Yih, Y.: A neural network integrated decision support system for condition-based optimal predictive maintenance policy. IEEE Trans. Syst. Man Cybern. Part A Syst. Hum. **37**, 226–236 (2007). https://doi.org/10.1109/TSMCA.2006.886368
40. Choo, B.Y., Adams, S.C., Weiss, B.A., Marvel, J.A., Beling, P.A.: Adaptive multi-scale prognostics and health management for smart manufacturing systems. Int. J. Progn. Health Manag. **7**, 014 (2016)
41. Li, X., Ding, Q., Sun, J.Q.: Remaining useful life estimation in prognostics using deep convolution neural networks. Reliab. Eng. Syst. Saf. **172**, 1–11 (2018). https://doi.org/10.1016/j.ress.2017.11.021
42. Kant, G., Sangwan, K.S.: Predictive modeling for power consumption in machining using artificial intelligence techniques. Procedia CIRP. **26**, 403–407 (2015). https://doi.org/10.1016/j.procir.2014.07.072
43. Rubaiee, S., Yildirim, M.B.: An energy-aware multiobjective ant colony algorithm to minimize total completion time and energy cost on a single-machine preemptive scheduling. Comput. Ind. Eng. **127**, 240–252 (2019). https://doi.org/10.1016/j.cie.2018.12.020
44. Mourtzis, D.: Design of customised products and manufacturing networks: towards frugal innovation. Int. J. Comput. Integr. Manuf. **31**, 1161–1173 (2018). https://doi.org/10.1080/095 1192X.2018.1509131
45. Hochdörffer, J., Moser, E., Lanza, G., Arndt, T., Hochdoerffer, J., Peters, S.: Customer-Driven Planning and Control of Global Production Networks-Balancing Standardisation and Regionalisation (2014)
46. Chryssolouris, G., Mavrikios, D., Pappas, M., Xanthakis, E., Smparounis, K.: A web and virtual reality-based platform for collaborative product review and customisation. In: Collaborative Design and Planning for Digital Manufacturing, pp. 137–152. Springer, London (2009)
47. Zhu, L., Cui, Y., Takami, G., Kanokogi, H., Matsubara, T.: Scalable reinforcement learning for plant-wide control of vinyl acetate monomer process. Control Eng. Pract. **97**, 104331 (2020). https://doi.org/10.1016/J.CONENGPRAC.2020.104331
48. Makris, S.: Cooperating Robots for Flexible Manufacturing (2021)

Chapter 2
Artificial Intelligence in Manufacturing Processes

Abstract A manufacturing process is defined as the use of one or more physical mechanisms to transform the shape and/or form and/or properties of a material. This chapter discusses AI topics, related to manufacturing processes. Initially, there is a short introduction to the main categories of manufacturing processes, namely, forming, deforming, removing, joining and material properties modification processes. Then, the chapter discusses the role of AI in supporting key activities, at process level, including (i) process monitoring and data processing, (ii) process modeling, optimization and control, (iii) fault diagnosis, tool wear prediction and remaining useful life estimation and (iv) process quality assessment and prediction. For each topic, the scope and the theoretical background are initially provided and then selected cases of AI applications are discussed. At the end of this chapter, both the impact and the limitations of AI at manufacturing process level are discussed.

Keywords Process optimization · Process control · Process monitoring · Adaptive control · Fault diagnosis · Data preprocessing · Alarms management · Tool wear

2.1 Introduction in Manufacturing Processes

"*A manufacturing process is defined as the use of one or more physical mechanisms to transform the shape and/or form and/or properties of a material*" [1]. A large variety of processes have been widely used to produce products, which in turn, has led to several proposals for the classification of processes. A high-level classification divides processes into *discrete parts* processes (where single items are processed) and *continuous processes* (where the material that is processed is continuous matter e.g. liquid). For instance, the metalworking industry involves the production of several single items thus, it utilizes discrete component manufacturing, whereas chemical processing, which is also involved in the fiber-making industries, uses continuous processes. Swift and Booker (2013) categorize processes as casting, cutting, forming, and fabrication Swift and Booker [2]. Kalpakjian and Schmid (2006) classified processes into six subcategories of casting, machining, finishing, joining, sheet metal, polymer processing, and deformation processes [3].

© The Author(s), under exclusive license to Springer Nature Switzerland AG 2023
G. Chryssolouris et al., *A Perspective on Artificial Intelligence in Manufacturing,*
Studies in Systems, Decision and Control 436,
https://doi.org/10.1007/978-3-031-21828-6_2

Finally, Chryssolouris (2006) suggested the following taxonomy of manufacturing processes [1], where the following five categories were introduced:

1. *Forming or primary forming processes*—these processes refer to the creation of an original shape from a molten or gaseous state, or from solid particles of an undefined shape (e.g. casting, melt processing). Normally during this type of process, cohesion among particles is created.
2. *Deforming processes*—processes that alter the original shape of a solid to another shape leaving its mass or material composition intact (e.g. extrusion, forging). During this process, cohesion is maintained among particles.
3. *Removing processes*—processes where material removal occurs (e.g. turning, drilling, milling) cohesion among particles is destroyed.
4. *Joining processes*—processes that unite individual parts to make subassemblies or final products (e.g. welding, adhesive bonding, mechanical joining). Additive processes, such as filling and impregnating of workpieces, are also included in this category; cohesion among particles is increased.
5. *Material properties modification processes*—processes that purposely change the material properties of a workpiece to improve its characteristics based on the requirements of each application without changing its shape (e.g. anodizing, electroless plating).

The selection of the set of manufacturing processes for the production of the final product is a critical decision point that affects cost, production rate, part quality, and so forth. The discussed categories of manufacturing processes are applied to a wide range of engineering materials (metals, ceramics, polymers, and composites) each of which is more efficient to be produced via certain processes that are mainly affected by the following two factors that seem to be the most important ones (Fig. 2.1):

1. The lot size of the parts to be manufactured: small lot sizes require flexible processes capable of accommodating different geometric features, etc. (e.g., material removal processes). On the other hand, large lot sizes enable the amortization of the high tooling costs of the primary forming or deforming processes, and hence their utilization.
2. Physical properties of the material (i.e. melting point): The high melting temperatures of metals suggests that they are usually processed in solid form i.e., using material removal and deforming processes. Polymers and composites have a lower melting point, allowing the use of other processes such as forming, where the material is often in a liquid state; Nevertheless, secondary operations, e.g., grinding, are often required to comply with the desired dimensional accuracy and surface quality. Ceramics are usually brittle, which means that it is difficult to process this type of material in solid form with the use of conventional machining techniques. Consequently, primary forming processes are preferred for the creation of the basic shape of the workpiece, and secondary operations (usually machining) are used for the creation of the final shape and surface quality.

Even though empirical rules allow for the selection of suitable manufacturing processes, the demands of the contemporary market have stimulated investments

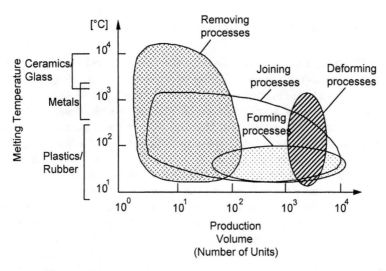

Fig. 2.1 The effect of melting point and lot size on manufacturing process application [1], Reprinted with permission from Springer-Verlag New York INC

in new technologies for manufacturing processes, including the innovative design of jigs and fixtures [4, 5]. Over the years, significant progress has been made by the improvement of hardware, energy sources, and the discovery of new materials. Additionally, both gradual and radical changes are expected by advanced manufacturing, which encompasses among other technologies, such as computer-controlled numerical control machines (CNC), automatic guided vehicle systems, computer-aided design (CAM), robotics, and rapid prototyping [6]. For instance, rapid prototyping, which is also known as additive manufacturing, 3D printing, and additive layer manufacturing, can allow for the transformation of freeform design to fully functional products, while promoting environmental advantages [7]. On the other hand, research on jigs and fixtures seeks to enable flexible work-holding solutions that will support the gradually increasing complexity of part sizes, shapes, machines, assembly, and cutting tools, while minimizing the required setup times [8, 9].

The following paragraphs will focus on the advancements that the emerging ICT technologies in tandem with AI can bring towards ameliorating the current practices in manufacturing processes. To this effect, a wide range of indicative applications focus on (a) process data collection, structuring, and fusion, (b) the adaptation of process models and the use of the Digital Twin concept for improved decision-making, (b) process quality assessment and prediction, (c) process optimization and control, and (d) tool condition monitoring and wear prediction are discussed.

2.2 Process Monitoring and Data Processing

The backbone of AI-based tools is the collection of meaningful data that will subsequently enable the development of intelligent solutions for more complex problems, namely process control. Under this prism, the first paragraphs of this section are dedicated to the use of AI for the collection of data in an automated, reliable, and effective manner and for making sense out of big, multi-sensor, multi-stream data collected at high sampling frequencies. Relevant subjects that are likewise discussed, include typical sensor systems installed in the industry in current practice, approaches that have been investigated for the exploitation of multi-source and heterogenous data, as well as AI-enhanced approaches that have been proposed for monitoring process parameters, including tool position and speed, temperature, pressure, current, etc.

With increasing automation, more and more human inspection activities are allocated to intelligent sensors, which have sensing and processing capabilities at the same or a superior level. By way of illustration, machine vision has been used to watch for product surface changes and tool breakage and touch-probes have eliminated the need to establish datum points among incoming material stock, cutting tools, and fixtures, whereas thermocouples and/or chemical sensors can provide information on whether the tool temperature is too high, which was previously detected by operators, watching for changes in the chip color and/or the smell of cutting fluids. Accelerometers and dynamometers can detect the excessive vibrations between tools and parts or even enable the anticipation of tool failure, which was previously achieved by the hearing sense of operators. Moreover, excessive cutting forces can be detected by strain gages and dynamometers, in place of the touching fixtures of tool holders.

As it occurs from the examples discussed, depending on the case, it might be reasonable for the sensors to monitor the tools, the workpieces, but also the performance of machines, given that with the use of the appropriate methodologies functionalities, such as the automatic process control, the supervision and error recovery actions can be enabled. Thus, sensors can be categorized into three groups, based on their function:

1. *tool monitoring*: this function enables among others the performance of (i) tool identification (e.g., bar code sensors), (ii) tool grip confirmation (e.g., proximity switches and limit switches), (iii) automatic tool change (ATC) and tool magazine operation confirmation (e.g., using proximity switches and limit switches), (iv) tooltip position confirmation (e.g., mechanical contact sensors), and (v) tool condition monitoring (e.g. Load cells, proximity switches and acoustic emission sensors are used for lathes, acoustic emission, and touch sensors used in machining centers).
2. *workpiece monitoring*: involves (i) work identification (e.g. proximity and limit switches, mechanical touch sensors), (ii) workpiece dimensions (e.g. contacting stylus, electrostatic capacitance, ultrasonic and electromagnetic sensors), (iii) mounting position (e.g. proximity switches, touch sensors, and air sensors), (iv) confirmation of mounting/faulty mounting, and (e.g. air sensors, air micrometers,

pressure sensors, and touch sensors), (v) unsatisfactory chucking or loose chuck (e.g. proximity switches, pressure sensors and limit switches).

3. *machine performance monitoring*: this function aims to satisfy requirements in the fields of quality control and automation as well as to improve the product's competitiveness and more details are provided in the process control section.

AI approaches for sensing and monitoring have already been studied from the second upsurge period of AI (the 1980s) [10–12], however, there have been some changes since then. The rapid advancement of technology results in increasingly complex mechanical equipment, with an integrated multi-sensor monitoring system, which in turn, perplexes the detection, and identification of faulty states of machine tools [13]. On the other hand, Machine vision is widely adopted for workpiece monitoring, as it is deemed to significantly improve defect detection in terms of efficiency, reliability, and quality. As a result, there is a greater variety in the captured data, since for instance, machine vision may extend outside the visible band of the electromagnetic spectrum, capturing data ranging from radio waves to gamma rays [14]. The large amounts of data acquired at high speed (e.g. videos), in combination with the networking possibilities of I4.0 has inflated the adoption of big data perspectives. The development of AI-based tools to monitor multimode processes has attracted limited attention, especially in the field of discrete part manufacturing [15]. Although these tools may be used in order to enhance the management of the complexity, to make accurate and reliable decisions despite nuisance, the time-varying process settings, dynamic process conditions, and non-stationary patterns. Indicatively, Yang and Zhou (2015) have developed a method, based on neural networks, aiming to locate costly false alarms, due to signals' disturbances [16].

The variability and richness of the collected signals call for the use of feature extraction (data processing) and then some technique for the detection and/or recognition of defects and faults, since the direct recognition of patterns and the inference of the process condition are almost impossible [17]. AI can empower the anticipation of detecting process anomalies and defects, given as input, a fusion of multi-sensor data embedded into the processing machines. In this case, proprioceptive, such as encoders, pressure sensors, etc. whose main purpose is to monitor the machine's condition and the energy source is used to extract relevant information about the process and the occurrence of undesired events. The contribution of AI in this scenario is the intelligent fusion of data in order to determine the stability of the process and to foresee process anomalies that may affect the final quality of the part.

Intelligent monitoring systems usually include preprocessing that is responsible for the transformation of the input patterns into low-dimensional feature vectors for easier match and comparison [18]. Next, the feature vectors are used as the input of AI techniques that map the obtained information in the feature space to types of product defects and/or equipment faults. Indicative AI tools that have been used comprise mathematical optimization, classification, and probability-based, as well as statistical methods. Focusing on classifiers and statistical learning, the k-nearest neighbor algorithms [19], the Bayesian classifier [20], the support vector machine

(SVM) [21], and the artificial neural network (ANN) [22] are the most popular and deep learning approaches that have started gaining interest [23].

Feature extraction has been frequently addressed with signal-processing methods e.g. time-domain analysis (mean, variance, kurtosis estimation), frequency-domain analysis (fast Fourier transform and bi-spectrum analysis), time–frequency analysis (Hilbert-Huang transform, wavelet transform, wavelet packet transform, sparse decomposition, etc.). Traditionally, the features are manually extracted, based on previous knowledge and expertise [24]. Nevertheless, in many cases, it is difficult to identify what features should be extracted to feed the AI-based techniques, which invite the use of deep learning. Deep learning algorithms can automatically learn features at multiple levels of abstraction, thus endowing, the learning of complex functions mapping the input to the output directly from data [25].

Furthermore, data fusion and feature dimensionality reduction methods are employed to simplify big data handling. For instance, it is a common practice to apply k-NN in combination with different reduction methods, such as principal component analysis or contribution analysis [26]. Chryssolouris et al. (1992) worked on the synthesis of multi-sensor information, using neural networks to synthesize the state variable estimates, based on the input of different sensors, as well as statistical criteria for the estimation of the best-synthesized state variable [10]. Deep convolutional neural networks performing multi-domain feature fusion were presented in [27]. Elbhbah and Sinha (2013) constructed a single composite spectrum, based on vibration measurements on multiple bearings to simplify the identification of faults in rotating machines [28].

Segreto and Teti (2019) deployed a multisensory monitoring system, comprising acoustic emission, strain, and current-based sensors, in a robot-assisted polishing of steel bars case study [29]. In their work, they present two alternatives for the extraction of relevant features, one using statistics, and another one using the wavelet packet transform. The features are used to construct different types of pattern feature vectors (basic and sensor fusion pattern vectors) that were propagated to a neural network pattern recognition algorithm in order to assess the roughness level of the surface of the polished part. Better success rates have been reported for the sensor fusion pattern feature vectors, which have confirmed the high efficacy of the sensor fusion technique in making full use of sensorial information.

C. Liu et al. (2018) proposed a machining condition recognition approach, based on multi-sensor fusion and support of the vector machine (SVM) in order to provide accurate recognition of machining conditions [30]. The suggested approach is based on a dynamometer sensor (collection of cutting forces) and an acceleration sensor (collection of vibration signals) processed with the wavelet decomposition method for the extraction of signal characteristics including means and variances. SVM is used as a condition recognition method, which is fed with one characteristic vector, where the signal characteristics of different sensors are extracted through the information fusion theory at the feature level. Zhang and Shin (2018) worked on turning process monitoring, using signals from two different types of sensors: a power meter and an accelerometer Zhang and Shin [31]. Their approach was based on signal processing and feature normalization to reduce dependencies on cutting conditions, workpiece materials, and cutting tools, as well as feature selection. The type-2 fuzzy basis function network is proposed for tool wear prediction, which is considered a regression problem, where the gradual growth of the tool wear needs to be monitored, whilst the chatter and tool chipping are dealt with as binary classification problems, via the SVM classifier. Grasso, Gallina, and Colosimo (2018) proposed the use of the one-class-classification SVM combining multi-sensor data into a multivariate space, where any occurrence of multi-sensor measurements, outside a region corresponding to controllable states, would lead to an automated alarm [32]. Wuest, Irgens, and Thoben (2014) worked with cluster analysis and Support Vector Machine (SVM) on product state data e.g. dimensions, surface roughness, etc. to describe the states of individual products along the entire process plan [33]. This approach has been proposed in place of traditional methods of modeling cause-effect relations to address the high complexity and high dimensionality of field data and improve quality monitoring.

Beyca et al. (2016) developed an approach, based on the Bayesian non-parametric Dirichlet Process for the real-time monitoring of ultra-precision machining process, by using multiple heterogeneous sensors, i.e. a miniature tri-axis force, tri-axis vibration and acoustic emission sensors, mounted in close proximity to the cutting tool [34]. A similar approach was proposed by Rao et al. (2015) who used thermocouples, accelerometers, an infrared temperature sensor and a real-time miniature video borescope in order to monitor additive manufacturing processes[35].

Physics-based approaches can also enable the processing of data for the identification of a process's status and whether the machines will retain the desired level of performance. Aivaliotis et al. (2021) presented the application of a method combining data having derived from multiple sensors integrated into the robot controller with the use of physics-based models in order to predict whether and when the robot will underperform [36]. This approach was based on [13], where a methodology was discussed for the selection of sensors to be used for the monitoring as well as the development of machine simulation to enable predictions for the remaining useful life of complex equipment.

2.3 Process Modeling, Optimization and Control

In traditional CNC systems, machining parameters are usually selected according to handbooks or previous experiences, leading to conservative configurations of process parameters for the avoidance of failures. Even if the machining parameters have been optimized offline by some optimization algorithm, their in-process adjustment is required due to tool wear, heat changes, and other disturbances that vary the system's dynamics. Thus, ensuring the quality of the products, reducing the machining costs, and increasing the machining efficiency, impose the online optimization and control of the machining process, i.e. the machining parameters must be adjusted in real-time to satisfy certain machining criteria.

The optimization of the process model parameters is an ongoing research subject. For instance, Mellal and Williams (2016) employed the cuckoo optimization algorithm and the hoopoe heuristic to optimize the control parameters of both conventional and advanced machining processes. Tamang and Chandrasekaran have proposed an integrated optimization methodology, using ANN, in combination with particle swarm optimization (PSO), for turning processes [38]. Rao, Rai, and Balic (2017) have created a new multi-objective optimization algorithm, having previously used and developed regression models as fitness function, tested on the wire-electro discharge machining process, the laser cutting process, the electrochemical machining process, and they focused on beam micro-milling [39]. Sadati, Chinnam, and Nezhad (2018) have developed a framework, which is agnostic of the optimization technique, with historical data for optimized process parameters design and process performance improvement [40].

The optimization of process parameters and process control strategies is usually achieved on the basis of process models. Process models are the descriptions that correlate the process variables measured by the sensors (e.g. cutting force) with the process parameters (e.g. feed and/or speed) so that the process controllers can achieve effective and efficient corrective actions. A prerequisite for the formation of the models is usually to know the objectives of the control strategy, together with the weighting of each objective compared with the others. Some researchers have deployed process Digital Twins (DTs), aiming to eliminate the manual data changes that are required in the digital models. The difference between the digital models and the digital twins is that digital models refer to the digital representation of the process [41], whereas a DT is a system that includes not only the digital model, but also the actual process as well as data and information flows between the two instances [42].

Given the need to accommodate environmental and social challenges, a large part of research on AI for adaptive control, has considered relevant criteria. AI techniques have been used for the prediction of power requirements in machining processes. Energy and production efficiency for machining process optimization is discussed in [43]. Dynamic prognosis is achieved by deploying a CNN on a fog node for the detection of potential faults, from the customized machining process through the evaluation of power signals, collected by power sensors. In this way, machines can be quickly stopped for maintenance in case of an abnormal situation. An energy-aware ant colony algorithm was used in [44] for the minimization of the total completion time and energy cost of a single machine.

Despite the complexity of manufacturing processes that emanates from non-linearities e.g. phase change, as well as the various physical phenomena that are usually involved per case e.g. self-oscillation, electromagnetic induction, etc., a wide range of process models (physics-based and data-driven) has been presented. Indicatively, the modeling and prediction of process forces in grind-hardening have been studied with a physics-based approach for the optimization of process parameters [45]. Finite element models have been also developed for laser drilling to determine the effect of the pulse repetition rate on temperature evolution and process efficiency [46]. The thermal modeling of the laser cladding process has been studied with a physics-based model for the prediction of the clad width and depth, given the processing speed and feed rate of the powder [47]. Pastras, Fysikopoulos, and Chrys-solouris (2017) studied the correlation of energy efficiency (required material volume, absorbed energy for material heating, process duration, etc.) with process variables (laser power, scanning speed) and the weld pool geometry (penetration depth, inter-face width, weld length, top and bottom surface concavity) in laser welding [48]. Finite element analysis and computational fluid dynamics were used to improve the velocity range of particles towards improving the adhesion quality in cold spray addi-tive manufacturing [49]. Wave propagation in cutting tools was studied by [50] with the use of Boundary Element Method 2D simulation, in order to get insights into the effects of tool geometry, machine vibrations, and noise in Acoustic Emission signals and improve control strategies that are based on such a kind of sensor feedback.

AI could be used to accelerate theoretical (physics-based) process models, towards their running in near-real time [51], where the term "near-real-time" in the context of decision making or optimization, refers to the ability to react to optimization requests within a fraction of process time. Figure 2.2 depicts an example of such functionality, where AI, with the help of mathematical physics, is used for the generation of a mesh for Finite differences that can improve the running time of the model itself. In another approach, Neural Networks have been employed for the production of an advanced simulation of grind-hardening [52], whilst [53] exploited the capabilities of data-driven methods for the derivation of process models to reduce the required domain knowledge that is required for the derivation of the counterpart theoretical models.

The automation of manufacturing processes is typically accompanied by the defi-nition and deployment of control loops which are responsible to tune process param-eters based on the target process results and feedback information about the ongoing status of the manufacturing process. Therefore, the automated mapping of field data to parameter adjustments such as the speed and feed, to obtain adequate opera-tion of the process invites the development of dedicated control systems. Indus-trial Control Systems is a common term, describing all types of control systems, comprising controllers (e.g. Programmable Logic Controllers), actuators, sensors, and other components, which are installed in an industrial facility for the collection of large-scale data in real-time.

The controller is typically an electronic hardware device that receives information from the monitoring system and utilizes the process model to compute the responses of the system's actuators. Proportional-Integral-Derivative (PID) controllers are among the most adopted controllers in industry thanks to their cost/benefit ratio [54]. The control schemes presented so far are of a wide variety, extending from relatively simple strategies, based on PIDs of fixed gains to more complex controllers, based on AI methods e.g. [55].

The motivation to seek advanced control strategies comes from the market demand for high reliability, high-precision, and high productivity of the manufacturing processes in order for the manufacturers to remain competitive. This in practice, means that the developed controllers should be capable of compensating for several sources of uncertainty, including even those arising from the errors/assumptions of the process models, based on which they compute the system's response. As an example, process models might induce errors or any uncertainty due to the dynamics

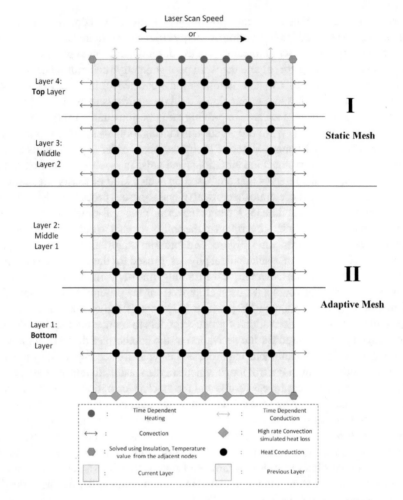

Fig. 2.2 Adaptive mesh for modeling, generated with the help of AI [51], Reprinted with permission from Elsevier

of the systems not being fully understood and the models not being as comprehensive as needed. In view of coping not only with process model errors, and uncertainties, but also with the varying dynamics of the systems performing manufacturing processes, the corresponding control systems require robustness and adaptivity to a certain degree [56].

Early controllers were usually based on fixed-gain controllers, which however, have limitations to addressing such conditions, due to the critical assumption in their definition that the system dynamics will not vary. Consequently, this type of a controller is heavily dependent on the accuracy of mathematical models and often ends up with deteriorating the system's performance, as the dynamics of physical processes gradually change [57]. In an effort to overcome the limitations of fixed-gain controllers, researchers have turned their attention to flexible controllers, which can seamlessly adapt to changes in signals and systems models. Chryssolouris, Domroese, and Zsoldos (1990) introduced the formulation of the process control as a decision-making problem of in-process selection of input variables, following the same logic as a human would, as opposed to contemporary ACO approaches, utilizing PID controllers [58]. This approach aims among others, to adapt to changes in the optimization criteria by changing the decision-making rules, as well as accounting for the effects of state variables, on the behavior of the overall process, at the price of higher in-process computational effort. Rojek and Kusiak [59] discussed control systems, based on data processing i.e. the analysis of registered data (including process parameters and signals) during the past production. Their approach searches, using a multi-agent methodology, the similarity among the current and the registered processes and selects the upcoming process parameters, according to the control function of the episode that was modeled via artificial networks.

Owing to this, several adaptive control systems have been proposed over the last years, which are defined as a class of controllers, capable of adjusting their own behavior, in response to process dynamics and disturbances [60]. The particularity of adaptive controllers, compared with other control schemes, lies in their adaptation mechanisms that may depend on initial knowledge, as well as on measurements (historical and/or ongoing). In more detail, adaptive controllers allow for the modification of the control law to overcome time-varying changes in any parameters of the process. Two kinds of adaptive control approaches can be distinguished. Firstly, the adaptive control constraint (ACC) operates by constraining the process within certain boundaries, in the attempt to maximize the machine tool and/or capabilities e.g., by constraining the maximum load for the prevention of machine overloading. The second kind is adaptive control optimization (ACO), which targets the optimization of the production process according to a set of optimization criteria. Indicatively, the prediction of the performance of adaptive control policies for phase change, in thermal-based processes, has been performed for conduction welding, solidification, and cooling down with empirical calibration of control laws, in combination with process modeling [61]. Commercially available systems mainly provide solutions with an integrated ACC approach, whilst ACO has not received much attention so far due to its requirements for elaborated mathematical process models and more sophisticated approaches to process monitoring [56].

With the advent of smart sensors and measuring technology [62] the access to power, force, acoustic emission, etc. data has been improved, whereas AI endows the ability to automatically apply corrective actions, while producing a part once a disruption has been detected [1]. Corrective actions may include the adjustment of process parameters in-process, or their adaptation within the process lifecycle, for the sake of minimizing or even eliminating the impact of anomalies on the achieved quality. In case it is not possible for corrective actions to be applied, early process interruptions are enabled allowing the prevention of time and resources waste.

Previous work on AI for adaptive control has addressed both the ACC and ACO approaches and uses techniques, such as Genetic Algorithms, Neural Networks, and expert rules. Liu, Zuo, and Wang (2010) presented a hybrid approach for milling processes, where ACO is achieved through a combination of neural networks and genetic algorithms, whereas ACC is based on neural networks and expert rules [63]. Dornheim and Link (2018) used AI, and in particular, multi-objective reinforcement learning, to enable reconfigurable ACO in manufacturing processes [64]. Their approach enables them to perform adaptive control when the relative importance of the control objectives is unknown during the design of the control, or when it changes, due to the varying production conditions and requirements. Moreira et al. (2019) developed a neuro-fuzzy logic model, which allowed for the implementation of a control strategy, for multiple machining variables control in a closed-loop, as well as real-time surface quality assurance in CNC machines [55].

The adaptive control concept has been frequently combined with the DT concept, i.e. enabling automated process parameters adaptation based on estimations, provided by digital models. Stavropoulos, Papacharalampopoulos, and Athanasopoulou (2020) worked on the hierarchical use of process models and the use of machine learning (variational autoencoder, adversarial networks) for the integration of simulation responses into a process DT, to make predictions and select process parameters towards meeting process requirements (Fig. 2.3) [65]. García-Díaz et al. (2018) introduced an OpenLMD architecture and a multimodal online monitoring system, which enables the use of machine learning techniques for process control and readjustments [66]. Actuators are hardware devices that accept the in-process control system commands as their input and convert these commands into adjustments to the process parameters. Papacharalampopoulos and Stavropoulos (2019) have developed a control-centric DT, based on a switched dynamic system model for thermal (laser-based) processes, which can be adapted, based on collected data, thus facilitating automated decision-making [67].

A DT was designed and implemented to manage uncertainty and robust process control in an additive manufacturing case study [68]. This approach is based on linear matrix inequalities (LMIs) for the control design and considers a DT lifecycle of a four-phase preparation, training, running, and generalization. The training phase, (Fig. 2.4), is crucial for the quantification of the material properties, whereas the integration of the uncertainty is achieved by defining a set of different material models that set the nominal process model, based on which the control can be designed. Further functionalities for the training procedure are (i) the modeling manipulation (retrieving real-time models out of theoretical models, which can then be integrated into the DT workflow for the formulation of the process control law) and (ii) the capability tracking (i.e. feasibility of the control form), which is required to ensure that the machine controller has the ability to assimilate a new control form. AI can manage the workflow of information and knowledge management similarly to [69]. An approach, based on the Dynamic Feature concept, has been proposed for the integration of machining, monitoring and the online inspection operations for optimized machining process control of complex parts [70].

Extending the concept of DT to the implementation of advanced human–machine interfaces, a DT architecture that enables users to remotely control and monitor the state of a fused deposition machine was presented by [71]. Another study has used

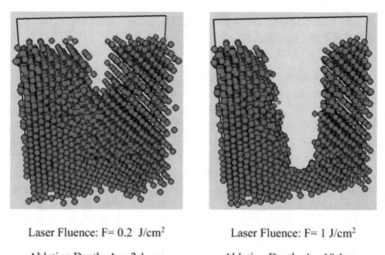

Laser Fluence: F= 0.2 J/cm^2 Laser Fluence: F= 1 J/cm^2

Ablation Depth: A_d= 2.1 nm   Ablation Depth: A_d=10.1 nm

Fig. 2.3 Molecular dynamics-based simulation [65], Reprinted with permission from Springer-Verlag London

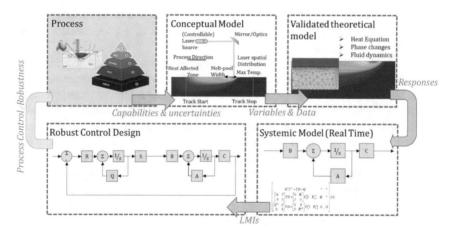

Fig. 2.4 Robust manufacturing implementation [68], Reprinted with permission from MDPI

Augmented Reality (AR), based on DT for process monitoring of laser-based manufacturing, enabling real-time information analysis and advanced data visualization of the process's performance discussed by [72]. Finally, the integration of robust control design with cyber security policies has been addressed with a game-theoretic approach [73].

2.4 Fault Diagnosis, Tool Wear Prediction, and Remaining Useful Life Estimation

Besides data processing and control, AI has attracted the attention of many researchers and seems promising in fault recognition applications. The recognition of the experienced faults can be regarded as a pattern recognition problem, where the power of AI algorithms has been frequently tested, proving their robustness and adaptation capabilities [25]. The motive behind developing solutions for fault diagnosis, but also the remaining useful life and tool wear prediction, includes among others, the reduction of impact on product quality, as well as the reduction or even the elimination of unexpected machine downtime. In particular, the objective of the developed methods is to diagnose and repair problems without human intervention.

Typically, tool condition monitoring and tool wear prediction are necessary to prevent problems due to physical phenomena that induce performance degradation in machine tools, such as excessive heat, produced by external or internal heat sources or disturbances in machining processes. For instance, a disturbance in the cutting process (because of a hard spot in the work material, for example) will cause a deflection of the structure, which may alter the undeformed chip thickness and in turn, alter the cutting force [1]. The initial vibration may be self-sustaining and cause the machine to oscillate in one of its natural modes of vibration, causing instability due to the regenerative effect, which is the dominant phenomenon, and/or the mode-coupling effect.

Overall, tool wear and fault detection methods can be classified into two approaches, direct (i.e. taking measurements of the tool itself to determine wear or breakage of the cutting edge) and indirect (i.e. using in-process measurements to monitor the tool condition) [1]. Direct methods for tool condition monitoring can be used to (i) check the dimensions of the workpiece; any deviation of the part geometry from the desired dimensions due to tool wear can be compensated for in subsequent cuts until the tool needs replacement, and (ii) check reflective properties of the surface, taking advantage of the fact that worn surfaces have higher reflective properties as compared with an unworn surface. The dirty conditions of the factory floor may undermine the performance of the required optics, (iii) check the tool's shape and geometry, using computer vision techniques [74].

Indirect methods for tool condition monitoring include (i) tool/work displacement methods, where the location of a machined surface is compared with the surface machined by an unworn tool. This approach is similar to the touch trigger direct method, but the direct method's disadvantages are avoided if the measurements are made in-process. (ii) electrical resistance methods, based on the varying area of contact between the tool and workpiece. The electrical resistance across the tool/workpiece junction increases as the tool wears. These methods may be influenced by the cyclic temperature variations from intermittent cutting operations. (iii) radioactive techniques, where a small amount of radioactive material is implanted in the flank face of the cutting tool at a known distance from the edge. When the material is removed, the tool is discarded. However, there are questions about the safety of these methods and the costs of clean-up and disposal. (iv) monitoring of the cutting forces by utilizing an analysis of the frequency components of the force or torque signal. Forces and torques are easily measured in-process, by using for instance, tool-holder dynamometers on lathes, table dynamometers on milling and drilling machines and machining centers, dynamometers built into the spindle bearings, or even by evaluating the cutting force, based on the spindle motor current, voltage, and speed. (v) A substantial amount of research has also been performed to develop methods of using acoustic emission (AE) signals for tool wear monitoring and tool breakage detection. (vi) Efforts have also been made to correlate the tool temperature with the tool wear. (vii) Several researchers have investigated the possibility of fusing the information from a variety of sensors to obtain more accurate and reliable estimates of the tool wear [56].

Direct methods lead to lost production time due to the inability of failure to be detected while the tool is being cut; thus, these methods may be acceptable for small lot sizes, where frequent access to the tool does not cost much production time [75]. Nevertheless, with the increasing industrial interest in in-line and in-process monitoring techniques, academics and practitioners focus more and more on indirect methods. Stavropoulos et al. (2014) investigated the indirect tool wear monitoring in milling processes and the isolation of vibration signals, due to tool wear, by excluding other vibration sources with process simulation [76].

The isolation of fault-relevant signals, as well as the detection and recognition of faults are the core underlying functions of a fault diagnosis technique [17]. This has to do with determining whether the equipment condition is normal or not, identifying the incipient failure and its causes, reasoning and predicting the trend of fault progression. AI approaches have the advantage of not requiring full prior physical knowledge, which may be difficult to be obtained in practice, explaining why a multitude of AI algorithms is used in this area, including k-NN, Naive Bayes classifier, SVM, and ANN [25].

Stavropoulos et al. (2016) investigated tool wear prediction on milling processes, using a simultaneous collection of acceleration and spindle drive current sensor signals, together with third-degree regression models and pattern recognition systems for their prediction [77]. Other approaches that investigated wear in milling operations, comprise the use of trend lines [78], neural networks [79] and random forests in place of ANNs and Support Vector Regression [80]. Neural networks have also been suggested for the tool wear prediction in drilling operations [81]. The arc-fault detection problem was studied in [82] by means of real-time deep neural networks, using Fourier coefficients, Mel-Frequency Cepstrum data and Wavelet features as input to differentiate normal from malicious current measurements.

Anagiannis, Nikolakis, and Alexopoulos (2020) performed an energy-based prognosis of the remaining useful life of the coating segments in hot rolling mill [83]. They analyzed segments of surface temperatures and hydraulic forces and used nonparametric statistical processes to predict the number of remaining products to be processed within a certain prediction horizon. The Maximum Likelihood Estimation was made in order for the probability of failure, within the defined prediction horizon to be assessed. However, common statistical process monitoring assumptions regarding the temporal dependence of process data are not often representative of the highly time-variant and non-stationary in nature manufacturing processes for complex products. Unstable and time-varying cutting conditions are caused by the repeated tuning and recalibration of process parameters in order for small lot size orders and highly personalized products to be served. This impacts the system's data acquired by the process monitoring, causing dynamic and non-stationary signals pattern. ANN algorithms are not suitable for highly varying scenarios, or they require a time-consuming training phase [84]. This is where deep learning solutions, e.g. Recurrent Neural Networks, can flourish, thanks to their capabilities to achieve high prediction accuracy, even with a limited training dataset and to cope with complex and time-varying process patterns. For instance, An enhanced deep autoencoder was implemented by [85] for the identification of faulty conditions in rotating machinery applications.

2.5 Process Quality Assessment and Prediction

The minimum variation in the process has been proposed by Genichi Taguchi as the definition of quality [86], whereas. Based on [1] *"The tolerance is defined as the range of values within which the particular dimensional characteristic of the product is acceptable"*. AI algorithms have the potential to detect patterns accurately, reliably, and quickly, which together with the advent of advanced inspection methodologies, namely X-ray computed tomography, enables the detection of the effects of process deviations and tolerances exceedance. In more detail, AI allows determining discrepancy models of both external and internal features and therefore has the potential to address a multitude of challenges, faced by quality engineers and practitioners.

Up until recently, statistical quality control has been one of the most popular approaches being based on control charts to monitor the output of manufacturing processes, identify unnatural variations in the measurements and specify their assignable causes [87]. The equipping of manufacturing systems with hundreds of sensors, though, has favored the collection of quality-related big data and the automation of analyses with the use of methods such as neural networks. Hence, the *Zero Defect Manufacturing* concept, which aims at decreasing and mitigating failures within manufacturing processes [88], has been provided with the required tools and data to flourish. The characteristic of this concept is that defects have to be prevented, and therefore, everything needs to be done in time scales, smaller than processing times, which differs from traditional quality control, i.e. six-sigma. In addition, corrective actions need to be identified for every single part being produced, as well as for every possible process anomaly that might arise.

An example of how AI can help process and quality engineers is the fast and robust in-line detection of process anomalies, for instance, by inspecting the product's surface. Neural Networks have been proposed for the estimation of the diameter of machined holes, in a metal removal monitoring application [89], as also surface roughness prediction [90, 91]. Statistics and the wavelet packet transform have been employed in a multi-sensory monitoring system, in combination with a neural network for the evaluation of polished parts' surface roughness [92]. Nacereddine, Goumeidane, and Ziou (2019) dealt with the automatic classification of weld defects, based on radiographic images. In particular, they have achieved the classification of defects into four different types (crack, lack of penetration, porosity, and solid inclusion) by using an unsupervised classifier, based on the multivariate generalized Gaussian distribution [93]. The same problem was addressed with a combination of PCA and SVM methods [94, 95]. PCA and an alarm rule, based on k-means clustering, are proposed for inspection in laser melting manufacturing, using frames of a video sequence [96]. The automated inspection in laser melting manufacturing, has also been the focus of Bugatti and Colosimo (2022), who compared the performance of an unsupervised k-means-based approach against Support Vector Machines and neural networks for the classification of hot-spots [97]. They have concluded that all tested classifiers were 80 times faster than the state-of-the-art PCA-based methods used for comparison.

Automating the surface inspection task in industrial applications, is challenging due to the costly data collection being a prerequisite to training appropriate methods, which tend to be highly dataset-dependent. Ren, Hung, and Tan (2018) proposed deep learning in an effort to alleviate these obstacles that were used for defect segmentation in three defect types [98]. In their approach, a classifier is built, based on the features of image patches, which are transferred from a pre-trained deep learning network, whilst pixel-wise prediction is obtained from the trained classifier over the input image. Deep learning has also been used by [99] for automatic visual inspection of defects such as scratches, burrs, and wears on surface parts. Imaging analysis with CNN (Convolution Neural Network) of training samples was run to determine whether defects exist in an image sub-region, whereas several types of deep networks of different depths and layer nodes, had been tested prior to concluding that a single CNN-based network is enough for the detection of several types of defects on textured and non-textured surfaces. Advantages in time and cost saving are reported over traditional manpower inspection systems.

The prediction of internal failures, before they are shipped, has attracted attention since it can prevent waste and increase customer satisfaction. The difficulty of this task falls into two categories, which are the scarcity of data, representing these failures and the limited ability of traditional machine learning algorithms for the optimization of non-convex metrics, such as Matthew's Correlation Coefficient (MCC). Maurya (2016) predicted internal failures with a meta-optimization algorithm that directly maximized MCC and a Gradient Boosting Machine (GBM) as a classifier for the meta-optimization algorithm [20]. Machine learning and sparse sensing were suggested in [100] for the prediction of shim gaps in aircraft assembly in order to reduce the need for gap filling, which is a time-consuming process causing production delays. This work assumes the existence of patterns, in shim distribution and performs a least squares regression for prediction of the shim occurrence.

A more complex problem is the quality prediction in the scenario of multistage machining processes, where various kinds of errors might influence the final quality, having derived from machine tools, cutting tools, previous machining stages, or even fixtures. Jiang et al. (2014) presented an approach, where error propagation networks have been employed to deal with the relationship among the consecutive processes [101]. The machining form-features and machining status were defined as the nodes of the network, back propagation neural networks were used so as to analyze the performance of the machining form-feature nodes, whereas support vector regression was used in order to predict the degradation trend of equipment.

2.6   AI Impact and Limitations

As a recap, AI-based tools can inflate the advancements in manufacturing processes and their impact can be captured and quantified via several key performance indicators, the most preferred among which are the following:

Cost reduction:

– The in-line qualification and monitoring have significantly reduced the costs by eliminating handling activities and waiting times for post-process inspections.

Footprint reduction:

– The in-line qualification and the monitoring have eliminated the need for separate post-process inspection workstations.

Scrap reduction & Higher machine availability:

– The timely detection of defects and/or underperformance of the processing tools, which in turn, triggers mitigation measures, such as process interruption or corrections, can critically reduce or even eliminate scrap (ZDM), as well as predict machine failures before they occur.

Productivity and Resilience:

– The AI-enhanced modules can promote efficient processing and process adaptations to different product volumes, materials, energy sources etc. by process parameters optimization, as well as adaptive control.

Nevertheless, there is still progress to be expected. The harsh environmental conditions (referring to the process's environment) sometimes restrict the use of specific sensors, thus, limiting access to potentially critical data for the development and deployment of AI based data. Additionally, the existence of legacy systems along with equipment, engineering, recurring, and training costs have prevented the integration of new sensors in industrial environments, in many cases.

Moreover, AI-based approaches may seem too complex to attract widespread adoption by practitioners. Important factors that may perplex AI-based applications comprise, but are not limited to (i) the type of process and process mechanisms that may introduce mechanical, thermal, or chemical interactions, (ii) the characteristics of the sensor, such as sampling rate, bandwidth, mounting possibilities, sensitivity to humidity, electromagnetic noise, etc., (iii) the desired granularity of analysis i.e.

the classification of quality monitoring outcomes may extend from accepted/not-accepted to a comprehensive set of potential process defects, which are difficult to manage (iv) the time frame required by the application (real-time process control, online quality monitoring, etc.). Furthermore, the lack of established models might be an additional obstacle. Indicatively, [102] claim that the lack of established quality or adaptive control systems for laser welding and laser cladding, has been the obstacle to exploiting data from the already integrated machines' thermal sensors for monitoring the temperature of the melt pool towards altering the process parameters to ensure process quality.

Additional limitations emanate from the required computational power to implement the AI-based tools. Several methods that have been presented, employ cloud resources for the required computational power. This may lead to processing considerable amounts of data, but it causes delays in data transferring, which undermines the effectiveness of the AI methods in process-level applications for which there are typical real-time requirements. In this context, Liang et al. (2019) have designed a fog and Convolutional Neural Network (CNN) enabled prognosis system for machining process optimization, which disentangles the requirements for local and intensive computations, hence, minimizing the bandwidth requirements [43]. Another limitation, concerning the methods that involve traditional machine learning techniques, is the requirement for the extraction of features, restricting the scalability and the performance of the solutions. Maggipinto et al. (2018) have proposed a deep learning approach, in the context of process optimization, aiming to provide estimations for quantities that are expensive or hard to measure [103].

References

1. Chryssolouris, G.: Manufacturing Systems: Theory and Practice. Springer (2006)
2. Swift, K.G., Booker, J.D.: Manufacturing Process Selection Handbook. Butterworth-Heinemann (2013)
3. Kalpakjian, S., Schmid, S.R.: Manufacturing Engineering & Technology. Pearson (2006)
4. Saleh, J.H., Mark, G., Jordan, N.C.: Flexibility: a multi-disciplinary literature review and a research agenda for designing flexible engineering systems. J. Eng. Des. 20, 307–323 (2009). https://doi.org/10.1080/09544820701870813
5. Beach, R., Muhlemann, A.P., Price, D.H., Paterson, A., Sharp, J.A.: Manufacturing operations and strategic flexibility: survey and cases. Int. J. Oper. Prod. Manag. (2000)
6. Maldonado, A., García, J.L., Alvarado, A., Balderrama, C.O.: A hierarchical fuzzy axiomatic design methodology for ergonomic compatibility evaluation of advanced manufacturing technology. Int. J. Adv. Manuf. Technol. 66, 171–186 (2013). https://doi.org/10.1007/s00170-012-4316-8
7. Bikas, H., Stavropoulos, P., Chryssolouris, G.: Additive manufacturing methods and modelling approaches: a critical review. Int. J. Adv. Manuf. Technol. 83(1), 389–405 (2015). https://doi.org/10.1007/S00170-015-7576-2
8. Yogeshkumar, K.S., Ramesh Babu, K.: An advanced method of jigs and fixtures planning by using cad methods. In: IET Conference Publications (2013). https://doi.org/10.1049/CP.2013.2565
9. Makris, S.: Cooperating Robots for Flexible Manufacturing (2021)

10. Chryssolouris, G., Domroese, M., Beaulieu, P.: Sensor synthesis for control of manufacturing processes. J. Manuf. Sci. Eng. Trans. ASME **114**, 158–174 (1992). https://doi.org/10.1115/1.2899768

11. Chryssolouris, G., Domroese, M.: An experimental study of strategies for integrating sensor information in machining. CIRP Ann. Manuf. Technol. **38**, 425–428 (1989). https://doi.org/10.1016/S0007-8506(07)62738-3

12. Chryssolouris, G.: Sensors in laser machining. CIRP Ann. Manuf. Technol. **43**, 513–519 (1994). https://doi.org/10.1016/S0007-8506(07)60497-1

13. Aivaliotis, P., Georgoulias, K., Arkouli, Z., Makris, S.: Methodology for enabling Digital Twin using advanced physics-based modelling in predictive maintenance. Procedia CIRP **81**, 417–422 (2019). https://doi.org/10.1016/j.procir.2019.03.072

14. Batchelor, B.G.: Machine Vision Handbook, pp. 1–2272 (2012). https://doi.org/10.1007/978-1-84996-169-1/COVER

15. Grasso, M., Colosimo, B.M., Semeraro, Q., Pacella, M.: A comparison study of distribution-free multivariate spc methods for multimode data. Qual. Reliab. Eng. Int. **31**, 75–96 (2015). https://doi.org/10.1002/QRE.1708

16. Yang, W.A., Zhou, W.: Autoregressive coefficient-invariant control chart pattern recognition in autocorrelated manufacturing processes using neural network ensemble. J. Intell. Manuf. **26**, 1161–1180 (2015). https://doi.org/10.1007/s10845-013-0847-6

17. Jardine, A.K.S., Lin, D., Banjevic, D.: A review on machinery diagnostics and prognostics implementing condition-based maintenance. Mech. Syst. Signal Process. **20**, 1483–1510 (2006). https://doi.org/10.1016/j.ymssp.2005.09.012

18. Yang, H., Mathew, J., Ma, L.: Fault diagnosis of rolling element bearings using basis pursuit. Mech. Syst. Signal Process. **19**, 341–356 (2005). https://doi.org/10.1016/J.YMSSP.2004.03.008

19. Wang, D.: K-nearest neighbors based methods for identification of different gear crack levels under different motor speeds and loads: Revisited. Mech. Syst. Signal Process. **70–71**, 201–208 (2016). https://doi.org/10.1016/J.YMSSP.2015.10.007

20. Maurya, A.: Bayesian optimization for predicting rare internal failures in manufacturing processes. In: Proceedings—2016 IEEE International Conference on Big Data, Big Data 2016, pp. 2036–2045. Institute of Electrical and Electronics Engineers Inc. (2016)

21. Delli, U., Chang, S.: Automated process monitoring in 3d printing using supervised machine learning. Procedia Manuf. **26**, 865–870 (2018). https://doi.org/10.1016/j.promfg.2018.07.111

22. Khorasani, A.M., Yazdi, M.R.S.: Development of a dynamic surface roughness monitoring system based on artificial neural networks (ANN) in milling operation. Int. J. Adv. Manuf. Technol. **93**(1), 141–151 (2015). https://doi.org/10.1007/S00170-015-7922-4

23. Lei, Y., Jia, F., Lin, J., Xing, S., Ding, S.X.: An intelligent fault diagnosis method using unsupervised feature learning towards mechanical big data. IEEE Trans. Ind. Electron. **63**, 3137–3147 (2016). https://doi.org/10.1109/TIE.2016.2519325

24. Alexopoulos, K., Nikolakis, N., Chryssolouris, G.: Digital twin-driven supervised machine learning for the development of artificial intelligence applications in manufacturing. **33**, 429–439 (2020). https://doi.org/10.1080/0951192X.2020.1747642

25. Liu, R., Yang, B., Zio, E., Chen, X.: Artificial intelligence for fault diagnosis of rotating machinery: a review. Mech. Syst. Signal Process. **108**, 33–47 (2018). https://doi.org/10.1016/j.ymssp.2018.02.016

26. Safizadeh, M.S., Latifi, S.K.: Using multi-sensor data fusion for vibration fault diagnosis of rolling element bearings by accelerometer and load cell. Inf. Fusion **18**, 1–8 (2014). https://doi.org/10.1016/J.INFFUS.2013.10.002

27. Huang, Z., Zhu, J., Lei, J., Li, X., Tian, F.: Tool wear predicting based on multi-domain feature fusion by deep convolutional neural network in milling operations. J. Intell. Manuf. **31**, 953–966 (2020). https://doi.org/10.1007/s10845-019-01488-7

28. Elbhbah, K., Sinha, J.K.: Vibration-based condition monitoring of rotating machines using a machine composite spectrum. J. Sound Vib. **332**, 2831–2845 (2013). https://doi.org/10.1016/J.JSV.2012.12.024

29. Segreto, T., Teti, R.: Machine learning for in-process end-point detection in robot-assisted polishing using multiple sensor monitoring. Int. J. Adv. Manuf. Technol. **103**, 4173–4187 (2019). https://doi.org/10.1007/s00170-019-03851-7

30. Liu, C., Li, Y., Zhou, G., Shen, W.: A sensor fusion and support vector machine based approach for recognition of complex machining conditions. J. Intell. Manuf. **29**, 1739–1752 (2018). https://doi.org/10.1007/s10845-016-1209-y

31. Zhang, B., Shin, Y.C.: A multimodal intelligent monitoring system for turning processes. J. Manuf. Process. **35**, 547–558 (2018). https://doi.org/10.1016/j.jmapro.2018.08.021

32. Grasso, M., Gallina, F., Colosimo, B.M.: Data fusion methods for statistical process monitoring and quality characterization in metal additive manufacturing. Procedia CIRP **75**, 103–107 (2018). https://doi.org/10.1016/J.PROCIR.2018.04.045

33. Wuest, T., Irgens, C., Thoben, K.D.: An approach to monitoring quality in manufacturing using supervised machine learning on product state data. In: Journal of Intelligent Manufacturing, pp. 1167–1180. Kluwer Academic Publishers (2014)

34. Beyca, O.F., Rao, P.K., Kong, Z., Bukkapatnam, S.T.S., Komanduri, R.: Heterogeneous sensor data fusion approach for real-time monitoring in ultraprecision machining (UPM) process using non-parametric Bayesian clustering and evidence theory. IEEE Trans. Autom. Sci. Eng. **13**, 1033–1044 (2016). https://doi.org/10.1109/TASE.2015.2447454

35. Rao, P.K., Liu, J., Roberson, D., Kong, Z., Williams, C.: Online real-time quality monitoring in additive manufacturing processes using heterogeneous sensors. J. Manuf. Sci. Eng. Trans. ASME **137** (2015). https://doi.org/10.1115/1.4029823

36. Aivaliotis, P., Arkouli, Z., Georgoulias, K., Makris, S.: Degradation curves integration in physics-based models: towards the predictive maintenance of industrial robots. Robot Comput. Integr. Manuf. **71**, 102177 (2021). https://doi.org/10.1016/j.rcim.2021.102177

37. Mellal, M.A., Williams, E.J.: Parameter optimization of advanced machining processes using cuckoo optimization algorithm and hoopoe heuristic. J. Intell. Manuf. **27**, 927–942 (2016). https://doi.org/10.1007/s10845-014-0925-4

38. Tamang, S.K., Chandrasekaran, M.: Integrated optimization methodology for intelligent machining of inconel 825 and its shop-floor application. J. Braz. Soc. Mech. Sci. Eng. **39**, 865–877 (2017). https://doi.org/10.1007/s40430-016-0570-2

39. Rao, R.V., Rai, D.P., Balic, J.: A multi-objective algorithm for optimization of modern machining processes. Eng. Appl. Artif. Intell. **61**, 103–125 (2017). https://doi.org/10.1016/j.engappai.2017.03.001

40. Sadati, N., Chinnam, R.B., Nezhad, M.Z.: Observational data-driven modeling and optimization of manufacturing processes. Expert Syst. Appl. **93**, 456–464 (2018). https://doi.org/10.1016/j.eswa.2017.10.028

41. Kritzinger, W., Karner, M., Traar, G., Henjes, J., Sihn, W.: Digital Twin in manufacturing: a categorical literature review and classification. IFAC-PapersOnLine. **51**, 1016–1022 (2018). https://doi.org/10.1016/j.ifacol.2018.08.474

42. Arkouli, Z., Aivaliotis, P., Makris, S.: Towards accurate robot modelling of flexible robotic manipulators. In: Procedia CIRP, pp. 497–501. Elsevier B.V. (2020)

43. Liang, Y.C., Li, W.D., Lu, X., Wang, S.: Fog computing and convolutional neural network enabled prognosis for machining process optimization. J. Manuf. Syst. **52**, 32–42 (2019). https://doi.org/10.1016/j.jmsy.2019.05.003

44. Rubaiee, S., Yildirim, M.B.: An energy-aware multiobjective ant colony algorithm to minimize total completion time and energy cost on a single-machine preemptive scheduling. Comput. Ind. Eng. **127**, 240–252 (2019). https://doi.org/10.1016/j.cie.2018.12.020

45. Salonitis, K., Tsoukantas, G., Stavropoulos, P., Stournaras, A., Chondros, T., Chryssolouris, G.: Process forces modeling in Grind-Hardening. In: Proceedings of the 9th CIRP International Workshop on Modeling of Machining Operations, pp. 295–302 (2006)

46. Stournaras, A., Salonitis, K., Stavropoulos, P., Chryssolouris, G.: Finite element thermal analysis of pulsed laser drilling process. In: Proceedings of the 10th CIRP International Workshop on Modeling of Machining Operations, pp. 549–553 (2007)

47. Salonitis, K., Stavropoulos, P., Stournaras, A. Chryssolouris, G.: Thermal modelling of the laser cladding process. In: Proceedings of the 5th LANE, pp. 825–835 (2007)
48. Pastras, G., Fysikopoulos, A., Chryssolouris, G.: A numerical approach to the energy efficiency of laser welding. Int. J. Adv. Manuf. Technol. **92**(1), 1243–1253 (2017). https://doi.org/10.1007/S00170-017-0187-3
49. Stavropoulos, P., Bikas, H., Bekiaris, T.: Combining process and machine modelling: a cold spray additive manufacturing case. Procedia CIRP **95**, 1015–1020 (2020). https://doi.org/10.1016/J.PROCIR.2021.01.178
50. Papacharalampopoulos, A., Stavropoulos, P., Doukas, C., Foteinopoulos, P., Chryssolouris, G.: Acoustic emission signal through turning tools: a computational study. Procedia CIRP **8**, 426–431 (2013). https://doi.org/10.1016/J.PROCIR.2013.06.128
51. Foteinopoulos, P., Papacharalampopoulos, A., Stavropoulos, P.: On thermal modeling of additive manufacturing processes. CIRP J. Manuf. Sci. Technol. **20**, 66–83 (2018). https://doi.org/10.1016/J.CIRPJ.2017.09.007
52. Tsirbas, K., Mourtzis, D., Zannis, S., Chryssolouris, G.: Grind-hardening modeling with the use of neural networks. In: AMST'99, pp. 197–206. Springer Vienna (1999)
53. Papacharalampopoulos, A.: Investigating data-driven systems as digital twins: numerical behavior of Ho–Kalman method for order estimation. Processes **8**, 431 (2020). https://doi.org/10.3390/PR8040431
54. Vilanova, R., Visioli, A.: PID control in the third millennium. Presented at (2012)
55. Moreira, L.C., Li, W.D., Lu, X., Fitzpatrick, M.E.: Supervision controller for real-time surface quality assurance in CNC machining using artificial intelligence. Comput. Ind. Eng. **127**, 158–168 (2019). https://doi.org/10.1016/j.cie.2018.12.016
56. Stavropoulos, P., Chantzis, D., Doukas, C., Papacharalampopoulos, A., Chryssolouris, G.: Monitoring and control of manufacturing processes: a review. In: Procedia CIRP, pp. 421–425. Elsevier B.V. (2013)
57. Anavatti, S.G., Santoso, F., Garratt, M.A.: Progress in adaptive control systems: past, present, and future. Presented at the August 11 (2016)
58. Chryssolouris, G., Domroese, M., Zsoldos, L.: A decision-making strategy for machining control. CIRP Ann. Manuf. Technol. **39**, 501–504 (1990). https://doi.org/10.1016/S0007-8506(07)61106-8
59. Rojek, G., Kusiak, J.: Industrial control system based on data processing. In: Lecture Notes in Computer Science (including subseries Lecture Notes in Artificial Intelligence and Lecture Notes in Bioinformatics), LNAI, vol. 7268, pp. 502–510 (2012). https://doi.org/10.1007/978-3-642-29350-4_60/COVER
60. Stefanovic, M., Safonov, M.G.: Safe adaptive switching control: stability and convergence. IEEE Trans. Autom. Control **53**, 2012–2021 (2008). https://doi.org/10.1109/TAC.2008.929395
61. Papacharalampopoulos, A., Stavropoulos, P., Stavridis, J.: Adaptive control of thermal processes: laser welding and additive manufacturing paradigms. Procedia CIRP **67**, 233–237 (2018). https://doi.org/10.1016/J.PROCIR.2017.12.205
62. Schütze, A., Helwig, N., Schneider, T.: Sensors 4.0—smart sensors and measurement technology enable Industry 4.0. J. Sens. Sens. Syst. **7**, 359–371 (2018). https://doi.org/10.5194/jsss-7-359-2018
63. Liu, Y., Zuo, L., Wang, C.: Intelligent adaptive control in milling processes. **12**, 453–460 (2010). https://doi.org/10.1080/095119299130182
64. Dornheim, J., Link, N.: Multiobjective reinforcement learning for reconfigurable adaptive optimal control of manufacturing processes. In: 2018 13th International Symposium on Electronics and Telecommunications, ISETC 2018—Conference Proceedings. Institute of Electrical and Electronics Engineers Inc. (2018)
65. Stavropoulos, P., Papacharalampopoulos, A., Athanasopoulou, L.: A molecular dynamics based digital twin for ultrafast laser material removal processes. Int. J. Adv. Manuf. Technol. **108**, 413–426 (2020). https://doi.org/10.1007/S00170-020-05387-7/FIGURES/14

66. García-Díaz, A., Panadeiro, V., Lodeiro, B., Rodríguez-Araújo, J., Stavridis, J., Papachar-alampopoulos, A., Stavropoulos, P.: OpenLMD, an open source middleware and toolkit for laser-based additive manufacturing of large metal parts. Robot Comput. Integr. Manuf. **53**, 153–161 (2018). https://doi.org/10.1016/J.RCIM.2018.04.006

67. Papacharalampopoulos, A., Stavridis, J., Stavropoulos, P.: Sensors performance in laser-based manufacturing process quality assessment: a conceptual framework. In: Procedia CIRP, pp. 490–494. Elsevier B.V. (2019)

68. Stavropoulos, P., Papacharalampopoulos, A., Michail, C.K., Chryssolouris, G.: Robust addi-tive manufacturing performance through a control oriented digital twin. Metals **11**, 708 (2021). https://doi.org/10.3390/MET11050708

69. Tao, F., Zhang, M., Liu, Y., Nee, A.Y.C.: Digital twin driven prognostics and health manage-ment for complex equipment. CIRP Ann. **67**, 169–172 (2018). https://doi.org/10.1016/j.cirp.2018.04.055

70. Li, Y., Liu, C., Gao, J.X., Shen, W.: An integrated feature-based dynamic control system for on-line machining, inspection and monitoring. Integr. Comput. Aided Eng. **22**, 187–200 (2015). https://doi.org/10.3233/ICA-150483

71. Mourtzis, D., Togias, T., Angelopoulos, J., Stavropoulos, P.: A Digital Twin architecture for monitoring and optimization of fused deposition modeling processes. Procedia CIRP **103**, 97–102 (2021). https://doi.org/10.1016/J.PROCIR.2021.10.015

72. Stavropoulos, P., Papacharalampopoulos, A., Siatras, V., Mourtzis, D.: An AR based Digital Twin for laser based manufacturing process monitoring. Procedia CIRP **102**, 258–263 (2021). https://doi.org/10.1016/J.PROCIR.2021.09.044

73. Zhu, Q., Başar, T.: A dynamic game-theoretic approach to resilient control system design for cascading failures. In: HiCoNS'12—Proceedings of the 1st ACM International Conference on High Confidence Networked Systems, pp. 41–46. ACM Press, New York, New York, USA (2012)

74. Stavropoulos, P., Salonitis, A., Stournaras, A., Pandremenos, J., Paralikas, J., Chryssolouris, G.: Tool condition monitoring in micro-milling—a critical review. In: Proceedings of the 5th International Conference on Manufacturing Research (2007)

75. Stavropoulos, P., Salonitis, K., Stournaras, A. Pandremenos, J., Paralikas, J., Chryssolouris, G.: Advances and challenges for tool condition monitoring in micro-milling. In: Proceedings of the IFAC Workshop on Manufacturing Modelling (2007)

76. Stavropoulos, P., Papacharalampopoulos, A., Souflas, T.: Indirect online tool wear monitoring and model-based identification of process-related signal. **12**, 1–12 (2014). https://doi.org/10.1177/1687814020919209

77. Stavropoulos, P., Papacharalampopoulos, A., Vasiliadis, E., Chryssolouris, G.: Tool wear predictability estimation in milling based on multi-sensorial data. Int. J. Adv. Manuf. Technol. **82**(1), 509–521 (2015). https://doi.org/10.1007/S00170-015-7317-6

78. Doukas, C., Stavropoulos, P., Papacharalampopoulos, A., Foteinopoulos, P., Vasiliadis, E., Chryssolouris, G.: On the estimation of tool-wear for milling operations based on multisensorial data. In: Procedia CIRP, pp 415–420. Elsevier B.V. (2013)

79. Drouillet, C., Karandikar, J., Nath, C., Journeaux, A.C., El Mansori, M., Kurfess, T.: Tool life predictions in milling using spindle power with the neural network technique. J. Manuf. Process. **22**, 161–168 (2016). https://doi.org/10.1016/j.jmapro.2016.03.010

80. Wu, D., Jennings, C., Terpenny, J., Gao, R.X., Kumara, S.: A comparative study on machine learning algorithms for smart manufacturing: tool wear prediction using random forests. J. Manuf. Sci. Eng. Tran. ASME. **139** (2017). https://doi.org/10.1115/1.4036350

81. Corne, R., Nath, C., Mansori, M.E., Kurfess, T.: Enhancing spindle power data application with neural network for real-time tool wear/breakage prediction during Inconel drilling. Procedia Manuf. **5**, 1–14 (2016). https://doi.org/10.1016/j.promfg.2016.08.004

82. Siegel, J.E., Pratt, S., Sun, Y., Sarma, S.E.: Real-time deep neural networks for internet-enabled arc-fault detection. Eng. Appl. Artif. Intell. **74**, 35–42 (2018). https://doi.org/10.1016/J.ENGAPPAI.2018.05.009

83. Anagiannis, I., Nikolakis, N., Alexopoulos, K.: Energy-based prognosis of the remaining useful life of the coating segments in hot rolling mill. Appl. Sci. **10**, 6827 (2020). https://doi.org/10.3390/APP10196827

84. Serin, G., Sener, B., Ozbayoglu, A.M., Unver, H.O.: Review of tool condition monitoring in machining and opportunities for deep learning. Int. J. Adv. Manuf. Technol. **109**, 953–974 (2020). https://doi.org/10.1007/S00170-020-05449-W/FIGURES/28

85. Ren, L., Sun, Y., Cui, J., Zhang, L.: Bearing remaining useful life prediction based on deep autoencoder and deep neural networks. J. Manuf. Syst. **48**, 71–77 (2018). https://doi.org/10.1016/j.jmsy.2018.04.008

86. Taguchi, G.: Introduction to Quality Engineering: Designing Quality into Products and Processes (1986). https://doi.org/10.2307/1268824

87. Powell, D., Magnanini, M.C., Colledani, M., Myklebust, O.: Advancing zero defect manufacturing: a state-of-the-art perspective and future research directions (2022)

88. Psarommatis, F., May, G., Dreyfus, P.A., Kiritsis, D.: Zero defect manufacturing: state-of-the-art review, shortcomings and future directions in research (2020)

89. Aguiar, P.R., Da Silva, R.B., Gerônimo, T.M., Franchin, M.N., Bianchi, E.C.: Estimating high precision hole diameters of aerospace alloys using artificial intelligence systems: a comparative analysis of different techniques. J. Braz. Soc. Mech. Sci. Eng. **39**, 127–153 (2017). https://doi.org/10.1007/s40430-016-0525-7

90. Karagiannis, S., Stavropoulos, P., Ziogas, C., Kechagias, J.: Prediction of surface roughness magnitude in computer numerical controlled end milling processes using neural networks, by considering a set of influence parameters: an aluminium alloy 5083 case study. Proc. Inst. Mech. Eng. B J. Eng. Manuf. **228**, 233–244 (2014). https://doi.org/10.1177/0954405413498582

91. Devarasiddappa, D., George, J., Chandrasekaran, M., Teyi, N.: Application of artificial intelligence approach in modeling surface quality of aerospace alloys in WEDM process. Procedia Technol. **25**, 1199–1208 (2016). https://doi.org/10.1016/j.protcy.2016.08.239

92. Segreto, T., Karam, S., Teti, R.: Signal processing and pattern recognition for surface roughness assessment in multiple sensor monitoring of robot-assisted polishing. Int. J. Adv. Manuf. Technol. **90**, 1023–1033 (2017). https://doi.org/10.1007/s00170-016-9463-x

93. Nacereddine, N., Goumeidane, A.B., Ziou, D.: Unsupervised weld defect classification in radiographic images using multivariate generalized Gaussian mixture model with exact computation of mean and shape parameters. Comput. Ind. **108**, 132–149 (2019). https://doi.org/10.1016/J.COMPIND.2019.02.010

94. Mu, W., Gao, J., Jiang, H., Wang, Z., Chen, F., Dang, C.: Automatic classification approach to weld defects based on PCA and SVM. Insight: Non-Destruct. Test. Cond. Monit. **55**, 535–539 (2013). https://doi.org/10.1784/INSI.2012.55.10.535

95. Stavridis, J., Papacharalampopoulos, A., Stavropoulos, P.: A cognitive approach for quality assessment in laser welding. In: Procedia CIRP, pp. 1542–1547. Elsevier B.V. (2018)

96. Colosimo, B.M., Grasso, M.: Spatially weighted PCA for monitoring video image data with application to additive manufacturing. **50**, 391–417 (2018). https://doi.org/10.1080/00224065.2018.1507563

97. Bugatti, M., Colosimo, B.M.: Towards real-time in-situ monitoring of hot-spot defects in L-PBF: a new classification-based method for fast video-imaging data analysis. J. Intell. Manuf. **33**, 293–309 (2022). https://doi.org/10.1007/S10845-021-01787-Y/TABLES/5

98. Ren, R., Hung, T., Tan, K.C.: A generic deep-learning-based approach for automated surface inspection. IEEE Trans. Cybern. **48**, 929–940 (2018). https://doi.org/10.1109/TCYB.2017.2668395

99. Park, J.K., Kwon, B.K., Park, J.H., Kang, D.J.: Machine learning-based imaging system for surface defect inspection. Int. J. Precis. Eng. Manuf. Green Technol. **3**(3), 303–310 (2016). https://doi.org/10.1007/S40684-016-0039-X

100. Manohar, K., Hogan, T., Buttrick, J., Banerjee, A.G., Kutz, J.N., Brunton, S.L.: Predicting shim gaps in aircraft assembly with machine learning and sparse sensing. J. Manuf. Syst. **48**, 87–95 (2018). https://doi.org/10.1016/J.JMSY.2018.01.011

101. Jiang, P., Jia, F., Wang, Y., Zheng, M.: Real-time quality monitoring and predicting model based on error propagation networks for multistage machining processes. J. Intell. Manuf. **25**, 521–538 (2014). https://doi.org/10.1007/s10845-012-0703-0
102. Purtonen, T., Kalliosaari, A., Salminen, A.: Monitoring and adaptive control of laser processes. In: Physics Procedia, pp. 1218–1231. Elsevier (2014)
103. Maggipinto, M., Terzi, M., Masiero, C., Beghi, A., Susto, G.A.: A computer vision-inspired deep learning architecture for virtual metrology modeling with 2-dimensional data. IEEE Trans. Semicond. Manuf. **31**, 376–384 (2018). https://doi.org/10.1109/TSM.2018.2849206

Chapter 3
Artificial Intelligence in Manufacturing Equipment, Automation, and Robots

Abstract The machines that perform manufacturing processes are the embodiment of these processes. This chapter discusses AI topics related to manufacturing equipment, automation, and robots. There is a wide range of problems requiring decision-making at the machine level, including the selection of the most appropriate type of machine for processing one or more parts, the identification of parameters for the machine's kinematic and dynamic models, the calculation of the optimum tool path, the selection of control strategy and gains. More specifically, the chapter examines (i) AI for manufacturing equipment definition—design—selection, (ii) task planning and machine programming, (iii) machine control and workstation orchestration and (iv) machine perception. For each topic the scope and the theoretical background is initially provided and then selected cases of AI applications are discussed. This chapter takes a deep dive into AI solutions for industrial robotics that are being widely used since their introduction in the manufacturing systems in the second half of the twentieth century.

Keywords Task planning · Machine programming · Workstation orchestration · Machine perception · Equipment selection · Equipment design

3.1 Introduction in the Manufacturing Equipment

"The machines that perform manufacturing processes are the embodiment of these processes. There is a close relationship between machines and processes since the capabilities and limitations of a process often depend on the design and operation of the machine performing it". "machines are used in manufacturing systems since the machines are the physical building blocks of these systems" [1].

Within this section, the analysis of the added value of AI will focus on one level higher in the hierarchy model that was presented in the introduction chapter (Fig. 1.6). This section discusses AI topics for manufacturing equipment, automation, and robots. There is a wide range of problems requiring decision-making at the machine level, including the selection of the most appropriate type of machine for processing one or more parts, the identification of parameters for the machine's kinematic and dynamic models, the calculation of the optimum tool path, the selection of control strategy and gains, e.g. the gains of Proportional-Integral-Derivative (PID) controllers, and many more. This section looks more closely into AI solutions for industrial robotics that are being widely used since their introduction in manufacturing systems in the second half of the twentieth century.

3.2   Manufacturing Equipment Definition—Design—Selection

The term "manufacturing equipment" covers a large variety of devices and machines, ranging from hand tools to complex automated machining centers, encountered in production systems. A machine tool can be defined as a non-portable machine with an integral power source, which causes the relative motion of a tool and a work-piece to produce a predetermined geometric form or shape. The discussion of the applications of AI for machinery equipment-related topics starts with the support for decision-making on machinery equipment selection. For equipment to be successfully integrated into the production system, several criteria should be met related to cost, flexibility, quality, ergonomics, and so forth, which can be highly impacted by the desired degree of automation. In modern manufacturing systems, the customer expectations necessitate flawless products, which in the level of manufacturing equipment can be analyzed into two categories of characteristics: (a) accuracy in geometric and kinematic terms, under static, dynamic, or thermal loading, and (b) reliability, i.e., high machinery equipment performance over an extended period. Other factors that may determine the selection of a particular machine could be maintainability, work space accessibility, ergonomics, and safety. In general, one can say that automation with the use of machine tools should strike a balance between cost, quality (particularly reliability), and flexibility. In this paragraph common types of machine tools that are used in production will be discussed categorized into the two major classes of machines: deforming and removing, as well as some typical types of industrial robots aiming to provide a comprehensive overview of the machinery equipment selection problem.

Deforming machines

Deforming machines usually have two or more tool parts that they bring together by providing the necessary force, energy, and torque for the process while ensuring adequate guidance of the tools to deform the workpiece. Deforming machines are typically large, low-cost, and unsophisticated controls and they are used in the metal working industry. These machines can be classified into two groups according to the relative movement of the tool parts: machines with linear relative movement, and machines with non-linear relative movement [2]. Machines not belonging to either of these categories are usually considered special-purpose machines. Furthermore, machines for deforming can be divided into [3] the following types: (1) Energy constrained whose characteristic feature is the available energy, which is converted into work on the workpiece; the deforming process is completed when this energy conversion is over; typical examples of such types of machines include belt-operated drop hammers, screw presses, etc., (2) movement-constrained machines that have constrained stroke which can be either determined through the eccentricity of a cam e.g. eccentric presses or the geometry of a crank shaft e.g. crank presses, (3) force-constrained machines, where the exerted force is controlled; the most widely used deforming machine that is force-constrained is the hydraulic press, which also provides notably accurate force control throughout the deforming process.

Machine tools for material removal

Material removal machines are the most used equipment type in manufacturing systems including machines whose working is based on mechanical, thermal, electrochemical, and chemical material removal mechanisms. Machine tools for material removal comprise: (1) a frame supporting the relative motions between the tool and workpiece for the cutting action to occur (2) the main and secondary drives providing for the main cutting action and the relative motion between the tool and the workpiece; (3) auxiliary devices that provide coolants and other necessary functions for the machine, (4) controls coordinating the movement of axes so that the motions of the tool, workpiece, as well as the resulting cutting action, to accurately produce the target workpiece geometry. In common practice, they either have a single cutting edge, such as in turning and milling, or multiple cutting edges, such as in grinding. In turn, machines with single cutting edges can be classified, depending on their cutting motion, into rotational machines (e.g. lathes, drilling, and boring equipment) and translational machines (e.g. gantry or double-column planning machines, shaping, and broaching equipment). Other processes include grinding, honing, and lapping are used to achieve high precision and good surface quality. These processes are most commonly used in high-precision manufacturing, where quality and precision are more important than production rate.

Machine tool design

Machine tool design, in terms of overall structure and kinematics, is primarily dictated by the process that the machine performs, the range of workpiece sizes, the required accuracy, and the operation mode (manual, semi-manual, fully automated, etc.). Additional, requirements and attitudes of the machine end user can also include maintenance procedures, cost, and a host of other factors. The design of machine tools is a multi-disciplinary process including aspects from applied mechanics to ergonomics due to the complex electromechanical structures of the machines that include many interconnected devices. This unique level of complexity, coupled with strict requirements in terms of kinematic accuracy, static and dynamic behavior, etc., causes the machine tool design process to rely heavily on empirical knowledge and expertise. For instance, force–deflection characteristics of the machine tool joints are highly non-linear and generally difficult to be analyzed nevertheless it is needed to analyze them as the joining methods of the elements of a machine tool can have a high impact on the machine's performance. The dynamic loads, developed during the operation of the machine, propagate through the joints to the different elements of the machine tool. Since the joints can account for up to 90% of the total deflection of machine tool structures and joint deflections can downgrade the accuracy and surface quality of the produced parts, the design goal is to maximize the stiffness of the joints.

Typically, engineering analysis delivering exact solutions for the structure of the machine, the size of its components, etc. is not the case, as the analysis rather enables the identification of trends and the creation of a framework for machine design. In general, stand-alone machines are designed to accommodate a variety of conditions in terms of workpiece size, etc., while dedicated machines, such as transfer lines are designed for reliability rather than flexibility. Machine tool frames are usually modular structures, made up of several elements e.g. baseplates, columns, beds, and crossbeams, whose shape, size, and material primarily depend on the position and length of the machine's moving axes, accessibility, and safety of the working space, the direction and magnitude of the anticipated process forces and manufacturability and cost of the machine. An example of how AI can support machine design includes the work of Romeo et al. (2020) who proposed an innovative machine learning algorithm for the prediction of machine specification parameters [4]. Decision/Regression Tree, k-Nearest Neighbors, and Neighborhood Component Features selection were adopted to extract decisional information to recommend the most suitable technical choice for designers and technicians and enabled the development of a Design Support System. Other methodologies for the prediction or estimation of machine components specification data and parameters include model-based approaches and simulation tools (e.g., CAE) [5]. Nevertheless, predictions and estimations required data which in industrial practice are typically insufficient, especially when it comes to fault data because equipment run in a healthy state most of the time. The Digital Twin of machines can solve this problem by generating synthetic data. However, the fidelity of the digital model, which is a function of the model itself, and the values of its parameters, determines the quality of synthetic data [6].

Industrial Robots

The types of robots that are usually employed in the industry can be categorized into the following six main categories: articulated robots, cartesian robots, SCARA robots, delta robots, polar robots, and cylindrical robots. Each type is characterized by specific advantages and disadvantages and performs better for specific applications. For instance, articulated robots allow for a wide range of movement and high flexibility, and thus they are commonly used for welding, packaging, machine tending, and material handling. On the other hand, the compact design of cylindrical robots is exploited for tight workspaces. Another taxonomy can be based on the level of decisional autonomy that is achieved by the robot, which includes the programmed, teleoperated, supervised, collaborative, and autonomous robots. In general, robots have contributed to decreasing the prices of goods by increasing productivity, as well as improving the quality of labor, and producing a greater variety of products and services. Therefore, it is expected that the utilization of AI, in combination with both the advancements in sensor technology [7] and the robotic structures' design [8] and manufacturing, will proliferate the capabilities and benefits of industrial robots. The low-cost computer hardware has shifted the efforts of boosting the performance of robots from the amelioration of their electromechanics to the improvement of its computational intelligence e.g. models, control algorithms, and trajectory optimization algorithms [9], which in turn has stimulated extensive research in computational intelligence of robots. The fields that are expected to benefit the most from AI advancements are discussed in this section, emphasizing the implementation of machine learning, meaning the systems' ability to learn, decide, predict, adapt, and react to changes that improve from experience without being explicitly programmed.

Machine selection

Many companies utilize industrial equipment such as industrial manipulators, CNC machines, etc. to improve the performance of their manufacturing systems. However, this type of equipment becomes more and more complex and thus requires determining the most suitable alternative at the beginning of the cell's design. Relevant considerations as per the Health and Safety Executive (Britain's national regulator for workplace health and safety) are the working conditions, the user of the equipment, the purpose of use, ergonomics, spatial constraints including maintenance activities, the energy used, as well as the substances use for production [10]. The increase in the number of available solutions, as well as the multitude of conflicting criteria, challenge the decision-makers in the selection of the appropriate piece of equipment. This decision-making problem is usually addressed with multiple attribute decision-making techniques, which are typically based on subjective statements for the alternatives (different pieces of equipment) and the criteria (such as cost, repeatability, velocity, reach load, etc.). For instance, Chatterjee, Athawale, and Chakraborty (2010) compared the use of VIKOR and ELECTRE methods for the robot selection problem [11], and Parkan and Wu (1999) studied the OCRA, and TOPSIS methods [12]. These methods aim at determining a compromise solution, by weighing the opinions of decision-makers. Other typical methods that have been used for managing the selection process efficiently include statistical models, mathematical programming (e.g. [13] studied Dimensional Analysis, which is a non-parametric method, for the case of multiple inputs and outputs comparisons), Case-based Reasoning [14], Graphical Method together with a framework for an exhaustive database of robotic arms and coding system [15], digraph and matrix methods [16]. The opinion of the experts or decision-makers is usually accounted for in these approaches, which led to the use of the quality function deployment integrated with other methods e.g. [17] developed an integrated decision model which combined quality function deployment (QFD) with fuzzy linear regression. Fuzzy approaches have been frequently used to capture the opinion of experts which is frequently incomplete or vague [18]. Finally, Axiomatic Design principles have also been proposed to help the decision maker decide based on a systematic, and objective basis [19]. A limited number of researchers also dealt with providing methodologies to act as decision aids, meaning that they will enable decision-makers to perform their duties in a more comprehensive and structured manner when it comes to selecting manufacturing equipment. Apart from typical technical criteria, some researchers extended their views into considering criteria related to ergonomics and human factors, e.g. human skills and training compatibility, usability, and so forth. [20] worked on the measurement of the probability of the selected advanced manufacturing technology to satisfy the desirable ergonomic requirements based on linguistic terms reflecting the opinion of experts in fuzzy environments. This work was based on a multi-attribute fuzzy axiomatic approach and its scope has been to assist the management of complex problems involving considerable amounts of quantitative and qualitative attributes. Under this prism, it also provided a comprehensive list of ergonomic attributes, which can impact the selection of advanced manufacturing technology selection, together with a hierarchical structure for integrating them into the decision-making schemes of companies. Aly, Abbas, and Megahed (2010) suggested a method for the selection of a suitable robot to serve a group of CNC machines within a predefined space focusing on the optimization of the robot base location [21].

Workstation design

The design of robotic cells refers to resource positioning while accounting for several factors such as the available footprint, resource reachability, ergonomics, and safety. Especially in the case of human–robot collaborative cells, the factors that should be considered increase further, as indicated by Michalos et al. (2015) who provided a comprehensive list of design considerations for safe human–robot collaborative workplaces, where they highlight the aspect of safety that is captured in defining the robot control schemes, the corresponding sensors to trigger the responses of the control, as well as the corresponding interfaces to communicate the robot status to the operators nearby Michalos et al. [22]. The consideration of all these factors, as well as the corresponding evaluation of the tentative selections over the satisfaction of process specifications (e.g. parts' weight and characteristics, process sequence, etc.) and performance criteria (e.g. productivity, safety, operator satisfaction, etc.), is a challenging task that imposes the need for decision support tools.

Simulation has been frequently employed to facilitate the evaluation of the different scenarios considering several aspects including human factors, whereas a comprehensive review of methods and tools helping in the ergonomic evaluation of manufacturing workstations is provided by [23]. These tools can extend from digital human modeling and software tools allowing for the calculation of metrics expressing postural risk (e.g. rapid upper limb assessment tool-RULA), or biomechanics risks e.g. NIOSH equation, to immersive simulations and AI-supported assessments of risks. Indicatively, Michalos, Karvouniari, et al. (2018) suggested a framework for workplace design and evaluation that is based on a collaborative scheme among operators and engineers and does not require the setup of physical prototypes [24]. In this framework, operators are invited to perform their tasks in an immersive Virtual Reality (VR) simulation while data are recorded to calculate metrics in real-time (e.g. cycle time, covered distance) and provide the corresponding visualization (e.g. heatmaps) to the production engineers as a support tool for workplace evaluation and redesign. A fuzzy-based inference engine for the assessment of full-body postural evaluation and tested in a setup activity for iron case production [25]. Their approach has considered the most relevant full-body evaluation checklists together with the visual acquisition of data through thermography, whereas the data processing has been achieved with triangular fuzzy rules. An approach using ANNs for tracking moving objects is presented in [26], while an unsupervised semantics-based approach to sensor data segmentation, in real-time activity recognition, is discussed in [27]. Sensor-based reconstruction of human activity enhanced with the aid of meta-data has been presented by [28] as an attempt to overcome shortcomings of supervised learning methods for the collection of training sets of sensor events, but also of knowledge-based methods that require manual modeling effort.

Typically, the layout of a flexible manufacturing cell is influenced by the characteristics of the material handling system in the sense that the configuration of the cell should allow for efficient utilization of the material handling systems used to serve the machines. For instance, the layout of a cell served by an Automated Guided Vehicle (AGV) should consider the accessibility and required clearances for the AGV to operate. Some approaches that have been used for the facility layout problem comprise the construction and improvement algorithms, which neglect the load/unload points of machines and the function of the material handling system, as well as the Quadratic Assignment Problem, whose application on the machine layout problem has been criticized due to the variety of sizes of the machinery equipment [29].

Tubaileh (2014) discussed an approach for determining the machine layout in robotic cells, as well as the feasible robot configurations, based on kinematic constraints [30]. A non-linear optimization model has been used for this purpose, solved with the available Sequential Quadratic Programming algorithm of Matlab. A multi-criteria decision-making framework based on heuristics, along with analytical models and simulations have been proposed for the automatic design of hybrid layouts in [31]. The backbone of this approach is the generalized model of resources and human–robot tasks, whereas the framework accounts for the input of the designer, the generation of alternatives, and their evaluation before visualizing the layout in 3D, which is the final step (Fig. 3.1). The design of a multi-robot work cell layout has been treated as a nonlinear problem, implemented for the manufacturing of a fuselage panel [32].

3.3 Task Planning and Machine Programming

Having selected the machine(s) to be used, one prerequisite for their operating them in production cells is to plan their tasks. Planning can be categorized into two categories: (a) macro planning which refers to ordering tasks, (re-)configuring assembly work cells, and assigning resources, and (b) micro-planning which includes path, motion, and trajectory planning, work instructions generation, and process parameters selection [33].

With regards to macro planning for a working station, researchers and practitioners that work on task planning seek to synthesize feasible plans having taken into account the resources' capacities, spatial and temporal constraints, etc. to achieve

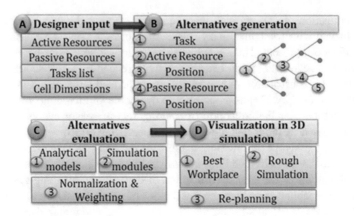

Fig. 3.1 Human–robot workplace design framework [31]

ad hoc objectives e.g., efficient manufacturing, high adaptability, and reduction of fixtures. This planning stage is especially challenging in the scheme of Human–Robot Collaboration, which is frequently an item of research as it is deemed to have the potential to keep the flexibility of modern industries at high levels thanks to the humans' intelligence and creative problem solving and the robots' repeatability, strength, and precision. A number of approaches for planning/coordination for HRI emphasizing the manufacturing/production environment were reviewed in [34]. For instance, Agostini, Torras, and Wörgötter (2011) proposed an approach for task planning in human–robot collaboration that is based on STRIPS (Standford Research Institute Problem Solver)-like planning operators and a competitive learning strategy to rapidly learn planning operators from few action experiences [35]. In this approach, many alternatives of cause-effect explanations are evaluated concurrently, and the most successful ones, as occurring from probabilistic estimations, in relation to other known estimates are used to generate the planning operators with enhanced confidence speeding up the learning. Takata and Hirano (2011) coped with task allocation by targeting the identification of the allocation pattern which has the highest potential to adapt to future changes by testing a set of possible allocations and trying to minimize the sum of total production costs which is estimated for a set of product change scenarios weighted by their occurrence probability [36]. Malvankar-Mehta and Mehta (2015) worked on the sharing of information that is required among a group of agents (humans or robots) following a nested optimization problem formulation, i.e. several hierarchy levels can have control over the decision variables, whilst the decision variable at one level may influence the objective function of other levels [37]. They considered a multi-level programming model level programming model of two levels (computer/agent level and leader-level) and a general form utility function to optimize the allocation of information, where the utility function at the agent level demonstrates the effectiveness optimization need, whereas at the leader-level represents the team performance optimization need.

Tsarouchi, Matthaiakis, et al. (2017) suggested an intelligent decision-making algorithm based on the depth search concept and the use of utility scores for multi-criteria evaluation for human–robot task allocation [38]. The proposed method enabled the allocation of sequential tasks assigned to a robot and a human in separate workspaces by considering process data, as well as data on the availability and characteristics of resources (Fig. 3.2) that were integrated into a Robot Operating System (ROS) framework. The focus was rather given to the human–robot coexistence for the execution of sequential tasks, to increase the automation level in manual or even hybrid assembly lines. Body gestures were the means of a human's interaction with a robot for commanding and guiding reasons. Another approach is the one in [39] where the use of a two-level hybrid hierarchical representation (resources and workload) into a level that can be translated into robotic commands is proposed. Additionally, an intelligent search-based multi-criteria decision-making algorithm was presented for task allocation in human–robot collaborative assembly applied in an automotive case study.

Alili et al. (2009) introduced the Human Aware Task Planner which is based on hierarchical task planning of two levels, i.e. operators (individual tasks), methods (a higher level task that can be decomposed into smaller ones), and a LAAS-based [40] architecture for autonomous systems in [41]. They extended the existing concepts

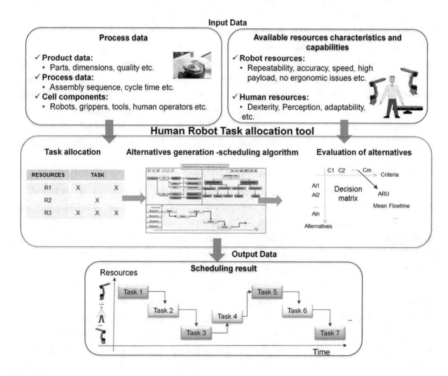

Fig. 3.2 Human–robot task allocation method overview [38]

by considering several synchronized streams dedicated to anticipating future potential actions of other agents and also considering social conventions and acceptable collaborative behaviors, intention explanation, situation analysis, and so forth. Chuan Tan et al. (2010) developed a framework based on task analysis to support human–robot collaboration planning, where qualitative and quantitative analysis of the tasks is accounted for to justify the possible collaborative solutions and detail the collaboration scenario [42]. Michalos, Spiliotopoulos, et al. (2018) introduced an approach for task planning based on CAD models for product assembly sequence extraction together with a heuristics-based search algorithm for the generation and examination of the assignment scenarios alternatives [43]. In this work, the output of the planner is the joint planning of task-to-resource assignment and cell layout, where the evaluation of each planning scenario is made against ergonomics, quality, and productivity criteria.

For a design to be manufactured, a set of process instructions, i.e. operation sheets or process plans, is necessary that describes the equipment, and/or people to be involved in the manufacturing process. Process planning bridges the gap between engineering design and manufacturing and establishes the sequence of the manufacturing processes to convert a part, from its initial into the final form. The process sequence incorporates process description, the parameters for the process, and equipment and/or machine tool selection. Process planning considers factors such as the shape and size of the workpiece, the required tolerances, the quantity to be made, etc., and requires the ability to interpret a particular design and substantial familiarity with manufacturing processes and equipment. Since often the designer's intention may not be obvious to the process planner; the designer is often unaware of potential manufacturing constraints and may produce a design that is either impracticable or costly to produce; the generation and execution of a production plan might be more time-consuming and involve several organizations distributed in different geographical locations; manual process planning is of limited consistency and optimization concerning certain performance criteria, computerized systems for process planning are needed [1].

Computer-aided process planning (CAPP) solutions usually fall into two major areas: *variant* process planning, where library retrieval procedures are applied for finding standard plans for similar components, and *generative* process planning, where plans are generated automatically for new components without reference to existing plans. *Variant process planning* is the easiest to implement. Variant systems allow for rapid generation of process plans, through the comparison of features with other known features in a database. However, to implement variant process planning, products should first be grouped into part families, based on feature commonality. Likewise, as the complexity of feature classifications increases, the number of part families also increases, causing excessive search times during the process of plan generation. Unlike the variant approach, which uses standardized process-grouped family plans, the generative approach attempts to imitate the process planner's thinking by applying the planner's decision-making logic and is based on defining the process planning logic with the use of methods such as decision trees, decision tables, artificial intelligence-based approaches, axiomatic approaches.

Generative process planning (GCAPP) relies on a knowledge base to generate process plans for a new design independent of the existing plans. The knowledge base is a set of rules derived from the experience of a human process planner. With generative methods, process plans can be generated for a wide variety of designs with dissimilar features. However, generative methods are difficult to be implemented in terms of constructing a set of rules, which can encompass all anticipated design features one is likely to encounter. There are three areas of concern in a generative process planning system: (a) component definition, i.e. the representation of the design in a precise manner in order to be made "understood" by the system, (b) identification, capture, and representation of the process planner's knowledge and the reasoning behind the different decisions made about the process selection, process sequence, etc., (c) compatibility of the component definition and the planner's logic with the system. The generative process planning can be executed either in a forward fashion (planning starts from the initial raw material and proceeds by building up the component using relevant processes), or backward, (planning starts from the final component and proceeds to the raw material shape). Two steps are frequently involved in this technique: part decomposition and feature recognition. AI can enable the analysis of a part's CAD drawing by using decomposition and feature recognition techniques so that a part's primary features to be identified. Many systems use backward chaining logic to generate and check the feasibility of a particular feature. Indicatively, Kardos et al. (2016) presented an approach for automated robotic assembly process planning, that includes feature-based models of assembly processes produced from standard CAD models of the products and the resources description, as well as the generation of constraints that ensure plan feasibility and the formal verification of fully specified plans [33].

An interest in automated planning, i.e. developing algorithms to decide the path or sequence of commands that a machine should execute to achieve a goal, has attracted research interest for years [44]. The goal states of automated tasks are associated with the features of the involved parts and their topology, as well as the process that should be executed as described in a part program. For machine tools, such as a milling machine or a lathe, the part program describes the path that the cutter will follow, the direction of rotation, and the travel rate. The increasing demand for new part programs has stimulated the development of CAD/CAM systems, such as the CADAM, CATIA, etc., that enable the generation of numerical-controlled (NC) part programs from CAD files, based on the geometric definition of a workpiece. The CAD/NC systems allow the user to rapidly define the geometry and use graphics display capabilities to quickly define, verify and edit the actual cutter motion preventing the use of valuable machine tool time. The computer can assist a part programmer by animating the entire tool path on the display terminal, showing the location of the cutter visually, and displaying the X, Y, and Z coordinates.

In addition, AI-based approaches can offer further capabilities, namely multi-axis and/or multi-head path planning. A Markov decision process is discussed in [45] for multi-head path planning for sheet metal manufacturing applications. The results support that multi-robot heads can be positioned more effectively by using the proposed algorithm, compared with the previous state. Ahmad and Plapper (2015) focused on the generation of safe trajectories for multi-axis CNC machines in non-functional trajectories, which has typically required the operator's intervention [46]. They developed an intelligent trajectory system based on image processing for background differentiation exploiting information from the CAM preparation and a vision camera, a search-based algorithm for generating trajectory points according to the machining strategy and avoiding the generation of safe points on obstacle envelope for collision avoidance in a virtual dynamic environment. The use of agents and NNs are discussed in [47] for acquiring and preparing the distribution of NC information to support NC planning. A combination of an agent-based organization and self-learning features, based on technological information is provided in order to support human engineers in planning and manufacturing. Energy efficiency is another criterion that can be critical for machine operation planning. Typically, the energy consumption models are either based on system models (which evaluate the energy consumption of a machine based on energy models of their sub-systems) or on process models (which focus on the relationship between material removal rate and energy consumption) [48]. Energy state is often coupled with many factors such as machine tool states, cutting, and tool conditions raising the difficulty in monitoring energy efficiency in machining. An expert system has been used together with Hidden Markov Models in [49] for energy state identification in milling processes.

Similar to machine tools robot programming also involves engineering tools and user interfaces that enable engineers to optimize the robot tasks, which are typically combined with manual programming. Online programming and offline programming are the two main categories of robot programming. Traditional methods for robot programming typically require either using the robot teach pendant or simulating the robot task inside a programming environment. The first case necessitates the training of the operators in properly using the teach pendant, and the inherent point-to-point programming style is efficient only for simple movements. In the second case which falls under offline programming, the process is based on models of the workstation and simulation of the robot, and hence requires financial investments for additional personnel and equipment. Also, knowledge of the platform-specific programming language (or of a 3D CAD program) is necessary, which is characterized by steep learning curves. Human is in the loop in this case, as it is frequently necessary to link activity-specific paths with each other to create the complete robot action sequence, but also to correct inaccuracies and errors that emanate from uncertainties. It is thus apparent that AI systems are required to decrease the demand on programming time, as well as system integration e.g. by enabling to automatically plan the robot motion given the initial and target status avoiding collisions, and visualize context relevant information to the human operator to support decision-making.

Some researchers have focused on the extraction of information from CAD models and the definition of the necessary semantic frameworks. The use of CAD models, with symbolic spatial relations, has been proposed for the automatic generation of assembly sequences and the reduction in requiring explicit robot programming [50]. In the same direction, an ontology-driven approach for automated software configuration has been developed [51]. Stenmark and Malec (2015) worked on a generic knowledge-based system architecture, where robot skills defined based on the product-process-resource triangle were used to allow the creation of composable robot tasks [52]. They used ontologies to represent the necessary knowledge for robot programming together with AI-based services that consumed the semantic descriptions of skills to help users instruct the robots. The approach was applied for the representation and execution of force-controlled tasks, using a natural language interface for the formation of preliminary tasks, "ABB RobotStudio" for detailed engineering, and a state machine for the task execution by one- and two-armed ABB robots in an industrial setting. Finally, the definition of languages for the planning problem specification has been an item of research interest. A broadly utilized formalism is the planning domain definition language—PDDL, which can be deemed as a transition system with Boolean facts as state variables [53].

Following the task planning, motion planning which involves searching for paths through the robot's configuration space to achieve moving a robot from an initial state to a target configuration while avoiding collision with objects in the surroundings is performed. Even though several robotized applications have been deployed in industry e.g. assembly, welding, painting, etc. their broader adoption is currently restricted due to the high engineering time required for the optimization and/or reconfiguration of robot trajectories [54]. When using conventional programming methods, robot path programming is performed by experienced robot programmers who need to spend considerable time programming the robotic paths which are application specific. In particular in the case of programming by demonstration, which is often used in industry, the robot is moved sequentially through several intermediate points that are recorded to the goal position using the teach pendant. Subsequently, the robot's path is generated by the robot controller that takes into consideration the dynamic constraints of the robot and interpolates the recorded points. For instance, Zöllner, Asfour, and Dillmann (2004) studied programming by demonstrating dual-arm humanoid robots. In this context, researchers and practitioners have investigated smart methods into enabling the automatic generation of the robots' paths [55]. The main functions to be boosted by AI in path planning, are the generation of alternative paths or robot configurations, along with the selection of the optimum path that would satisfy several criteria: for instance obstacle avoidance, shortest distance, etc.

The classification of motion planners is not an easy task however, it is common to refer to roadmap-based planners and tree-based planners [56]. Starting with the roadmap-based panners, Kavraki et al. (1996) introduced the probabilistic road-map path planning for computing collision-free paths for robots in a two-phase approach: (a) learning, where a probabilistic roadmap is constructed by generating the robot's free configurations randomly, and (b) query connecting the configurations using a local planner [57]. Bayazit, Lien, and Amato (2002) discussed several strategies for node generation and multi-stage connection strategies for cluttered 3-dimensional workspaces in the context of an obstacle probabilistic roadmap method [58]. Koo et al. (2011) discussed a sampling-based approach for deformable parts' manipulation [59]. Kohrt et al. (2013) proposed an online path planning and programming support system transforming the user's interaction into a simplified task that generates acceptable trajectories, applicable to industrial robots [60]. This work used a combination of Voronoi roadmaps including reachability, A* search, and elastic net trajectory generation.

The concept of Rapidly-exploring the Random Tree was presented by [61], and it is based on an initial sample being the root of the tree and newly produced samples which are then connected to the samples already existing in the tree. Tree-based planners have proven to be a good framework for dealing with real-time planning and re-planning problems. In [62] a re-planning algorithm for repairing Rapidly-exploring Random Trees when changes are made to the configuration space was presented, where the algorithm efficiently removes only the obsolete parts and maintains the rest of the tree. Heuristic beam search in combination with AR-based interfaces has been suggested for collision-free path generation and easy robot programming [63]. Wu et al. (2009) proposed a two-stage motion planner that consists of an offline stage and an online stage [64]. In the off-line stage, which is more time-consuming, the positions of the obstacles in the workspace are computed and stored using a hierarchical data structure with non-uniform 2 m trees. In the online stage, the real obstacles are identified and the corresponding 2 m trees from the pre-computed database are superposed to construct the real-time space. The collision-free path is then searched in this C-space by using the A* algorithm. Complex cluttered environments are considered in [65] and have been approached via random trees for optimal motion planning.

A method for trajectory optimization for multi-robot handling compliant parts, which can achieve collision avoidance and minimize deformations during manipulation, has been proposed in [66]. This method employs a non-linear programming model by exploiting a Response Surface Model based on FEM-generated data for optimization. Kaltsoukalas, Makris, and Chryssolouris (2015) introduced a grid-based search algorithm that uses hierarchical modeling of the robot configurations to gradually approach the target robot state by selecting and evaluating a number of alternatives [54]. The search algorithm makes use of heuristics for the generation of the robot configurations and its processing time can be adjusted to the user's requirements achieving to critically lessening the need for the recording of intermediate points. In this way, the proposed method can grant an inexperienced robot programmer the flexibility to generate automatically a robotic path that would fulfill the desired criteria, such as the shortest path. The alternative configurations are generated by emphasizing the robot's joints that determine to a higher degree the end effector position. The grid of the robot's alternative configurations can be adjusted by a set of parameters that affect the resolution of the grid and thus the search space. High grid resolutions result in smooth paths, whereas lower resolution minimizes computational time. Figure 3.3 depicts the main components of this approach, as well as an example of a set of alternatives that resulted in specific values of the following three user-defined parameters: (a) decision horizon (DH) that takes values from 1 to the n degrees of freedom of the robot, (b) the Maximum number of alternatives (MNA), and (c) Sample Rate (SR) which is defined as the number of samples taken from the joints, outside the decision horizon, to form the robot's complete alternative configurations. Motion planning can be particularly challenging in the case of flexible material manipulation, such as ropes, clothes, cables, etc.

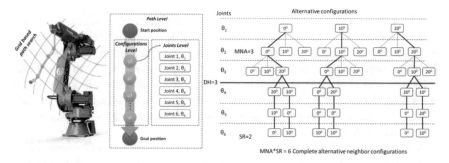

Fig. 3.3 Hierarchical levels for robot path planning and alternative robot configurations [54], Reprinted with permission from Pergamon

Additional planning problems that are relevant in automation and robotics include navigation among movable obstacles [67], pick-and-place planning [68], manipulation planning [69], and rearrangement planning [70], as well. The navigation of mobile robots includes modeling of their surroundings, localization of their position, motion control, obstacle detection, and avoidance, whereas the most critical function of navigational techniques is safe path planning. Mobile robot navigation has been categorized into global and local navigation, wherein the first case information about the environment, the goal position, and the position of the obstacles are required in contrast with the second that can deal with unknown environments. Known environments have been dealt with classic approaches such as cell decomposition, roadmap approach, and artificial potential field, whereas unknown or partially known with reactive approaches such as genetic algorithm, fuzzy logic, firefly algorithm, ant colony optimization, cuckoo search, etc. [67]. A q-learning algorithm is discussed in [71] for local path planning of mobile robots. Dynamic obstacle avoidance has been addressed with Polar Object Charts and supervised Machine Learning [72]. Furthermore, systems detecting collisions have been developed via Artificial Neural Networks [73] The artificial potential field method has been used for problems of high dimensionality, e.g. for problems involving multiple agents (robots, humans, AGVs, etc.), or problems where larger areas are covered by the robots. For instance, the safe movement of multiple mobile robots around humans [74], and the path planning of a cable parallel robot [75] are some of the problems falling into the discussed category of problems. However, there are reactive approaches to deal with similar problems as well, e.g. a hybrid particle swarm algorithm, combined with a gravitational search algorithm, is presented in [76] for multi-robot path planning concerning their energy consumption.

Makris, Kampourakis, and Andronas (2022) suggested an approach that allows robots to adjust their behavior so that the co-manipulation of fabrics between humans and robots is possible [77]. The authors focused on the interpretation of human manipulation actions i.e. perceiving the current postures and grasping points of humans, and their translation into robot reactions. In this scope, simulation based on the mass-spring model was employed to estimate the fabric's distortion. The main requirement in deformable object handling is the reliable control of deformation through efficient control of Grasping Points, thus the authors provide a set of strategies to plan the grasping point (GP) of the robot on the fabric given particular manipulation scenarios (human-one arm robot translation or rotation of fabric, human dual arm robot translation or rotation of fabric) the monitoring of human actions which can affect the generation of target grasping points. The approach has been validated for the translation and rotation in co-manipulation scenarios, in two use cases, inspired by the automotive composite industry (Fig. 3.4).

It is important to highlight that human–robot collaboration (HRC) comes with a cost, as it entails human safety-related challenges, e.g. collision of human and robot, and thus calls for dedicated mitigation measures. Starting with the discussion of the impact of these measures in planning, it is noted that the typical mitigation measure is for motion controllers to modify the speed of robot motions to improve safety when triggered by some sensor signal that monitors the human–robot distance [78]. This in turn can induce important variations in the robot task cycle time and increase the difficulty in planning for a long planning horizon, considering performance criteria, since robot trajectories are computed online based on the human position that may occur each time [79]. A common approach is to replan the robot tasks, based on the ongoing human position, which frequently penalizes the efficiency of the production

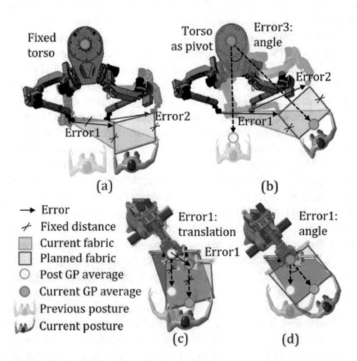

Fig. 3.4 Indicative handling strategies: **a** operator-dual arm robot translation, **b** operator-dual arm robot rotation, **c** operator-single arm robot translation, and **d** operator-single arm robot rotation with pivot axis at robot tool center point [77], Reprinted with permission from Edition Colibri AG

performance. AI has been used to exploit the HRC potential seeking among others to maximize the throughput by the proper task allocation and coordination of human and robot tasks, which at the same time imposes the need for managing temporal uncertainty while achieving robustness, flexibility, and reliability. Evangelou et al. (2020) distinguished the planning of operations in search-based and timeline approaches [80]. The work of the authors falls in the first category whereas the work of [81] falls into the second. Cesta, Orlandini, and Umbrico (2018) discussed an approach accounting for the description of production processes, operational constraints, relevant temporal features, and other factors together with a closed loop control architecture, where human is put in the loop and replanning is used to managing external events, such as a robot failure [81].

On the other hand, the tools that have been suggested to support the online robot programming, but also HRC require perception ability, as well. Since the implementation of human–robot interaction has been affected by the user's expectations [82], the design of the collaborative tasks should facilitate the monitoring of the robots' operations [83]. The tools that have been developed to support humans usually involve indirect or direct human–robot interaction (HRI). In the first case, AR-based interfaces, vision systems, wearables, and microphones are used, whereas in the second existing or imminent physical contact needs to be monitored with sensors such as force/torque or proximity. For instance, [84, 85] suggested Augmented Reality (AR) tools to visualize robot-related information in support of the human–robot interaction, Robot programming support tools are typically a combination of algorithms for optimizing motion planning, as well as intuitive interfaces for programming making use of gestures, visual features, color, shape, and contour-based approaches [86]. Voice commands are also popular in HRI; however, their effective implementation can be impeded in noisy industrial environments, unless filtering is used [87]. Hogreve et al. (2016) and Kaczmarek, Hogreve, and Tracht (2015) proposed the combination of gesture control, progress monitoring, and worker support [88, 89]. Intention awareness, which enabled the adaptation of the robot program, based on the user intention, has been tackled by converting robot programs into Markov chains and using position, gestures, and voice data [90]. Supportive AI has been identified as a critical factor for further advancements [91].

Robot programming methods such as walk-through programming, which is also referenced as "lead-through programming", or "manual guidance", have been proposed with early practical applications e.g. welding [92]. In these programming strategies, the human operator moves the robot end-effector and manually leads it to the target location, without needing any prior knowledge of the particular robot programming language and of the functionalities of the teach pendant. However, these approaches require effort into estimating the areas where the robot can move in order to minimize unwanted collisions. For example, Neto, Pires, and Moreira (2009) investigated robot programming with a combination of gestures, speech, and force control [93]. Gkournelos et al. (2018) discussed the use of wearables for operator support in human–robot collaborative industrial workplaces [94]. The application that was developed in their work supports natural language processing, automatic estimation of the working space where the robot is safe to move to avoid collisions,

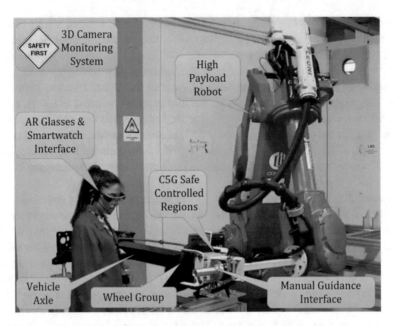

Fig. 3.5 Automotive hybrid assembly cell [94], Reprinted with permission from Elsevier BV

and context-aware visualization of data. The application has been deployed in a smartwatch for direct interaction with the robot via audio commands, and manual guidance, whilst supportive visual information can be provided by AR-based visualization systems. Figure 3.5 depicts the testing and validation of the application in an automotive case study.

Dimitropoulos et al. (2020) presented a comprehensive approach where voice commands and human action recognition were used to adjust the robot behavior toward improving the operator's experience [95]. Apostolopoulos et al. (2022) worked on a two-mode training framework (Fig. 3.6). The first mode entitled "training mode", focused on introducing the operator to hybrid workstations, whereas the second named "assistive mode", focused on supporting the assembly of new products by providing online instructions [96]. Both modes are based on a step-by-step walkthrough experience, at the same time, object recognition based on machine learning is used to provide augmented identification and instructions, based on the positioning of the physical objects and the robot. The object recognition algorithm takes into consideration the operator's head movements, the random positioning of components, and the dynamic environmental conditions. Furthermore, it is built with the use of the AR headset's vision sensor to ensure the easy setup of the application.

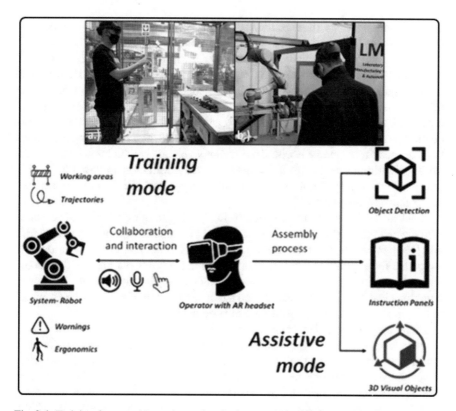

Fig. 3.6 Training framework's modes and main features [96], Reprinted with permission from Elsevier BV

3.4 Machine Control and Workstation Orchestration

As it occurs from the discussed examples, the contemporary production environments where hybrid production paradigms are deployed, dynamic updates of the machines' surroundings are held constantly advanced perception abilities are required to detect disturbances or even safety risks and trigger the required replanning actions. Moreover, it is common that the plans produced offline based on the simulated environments involve ideal conditions and deterministic scenarios, which is not the case in the real world. Therefore, schemes for execution control, motion control, force and torque control, and others are used to ensure that the planned goal is achieved each time. For instance, online correction of offline routines, through force and vision sensors, has been investigated in an automotive and an aeronautics case study [97]. In more detail, the adaptation of the robotic motion is based on force control to address variations in the position of the objects and the changing environment, as well as the fine-tuning of the force to be applied to the parts. The vision algorithms detect deviations during the execution of the robotized tasks, which lead either to the modification of the robotic path or notifying the human to resolve the issues.

A control algorithm, with the use of GA, is described in [98] for enabling a visual target tracker robot. As reported in the study, the application of the GA algorithm in the torque control of the robot has outperformed traditional tuning techniques in terms of mean square error, overshoot, and settling time. Adaptive speed control of a permanent synchronous motor has been achieved in [99] using a state feedback controller online, adapted by an artificial bee colony algorithm, improving the control performance, in comparison with a non-adaptive one. A combination of the evolutionary algorithm with swarm-based intelligence algorithms is proposed in [100] to tune a PID controller to reduce the time required by stochastics, naturally inspired and population-based algorithms.

Son (2014) worked on the division of intelligent jamming region with machine learning and fuzzy optimization for the control of a robot's part micro-manipulative task [101]. In particular, a rule-based learning algorithm and fuzzy optimization were suggested for controlling a robot, which was intended to mate parts with a target, without jamming by adjusting the parts' lateral and angular movements. This work comprises learning from experience and a reduced number of jamming areas, which in turn, enables the faster performance of assigned tasks.

Dexterous multi-fingered hands can enable robots to flexibly perform a wide range of manipulation skills. However, many of the more complex behaviors are also notoriously difficult to control: Performing in-hand object manipulation, executing finger gaits to move objects, and exhibiting precise fine motor skills, such as writing, all require a delicate balancing of contact forces, breaking and reestablishing contacts repeatedly, and maintaining control of underactuated objects. In [102] a method of online planning with deep dynamics models (PDDM) that addresses both of these limitations is demonstrated. This study shows that AI advances can facilitate the training of complex behaviors directly with real-world experience on physical hardware, precluding the need for sim-to-real transfer or prior system/environment-specific information in general.

The manipulation of fabrics may require not only repeatability, and precision, but also involves repetitive movements and the assumption of incongruous postures, which typically would raise the need for automation. Nevertheless, flexible material deformation limits the robot's cognition during fabric handling. A model-based closed-loop control framework that allows human–robot or multi-robot fabric manipulation has been proposed in [103] towards addressing these challenges. A mass-spring model was used to simulate ply distortion and generate the spatial localization of the optimal grasping points (GPs). The model is enriched with the operator's real-time handling actions captured by a perception system (Fig. 3.7). The proposed sensor and model-based controlling framework incorporate robot motion planners, accounting for non-rigid object human–robot co-manipulation, or synchronization of cooperative robots within fully automated tasks.

Autonomous task orchestration is discussed in [104]. An HMM is developed for modeling the autonomous task orchestration, while a state machine logic handles manufacturing exceptions as well as the determination of batch sizes in an adaptive

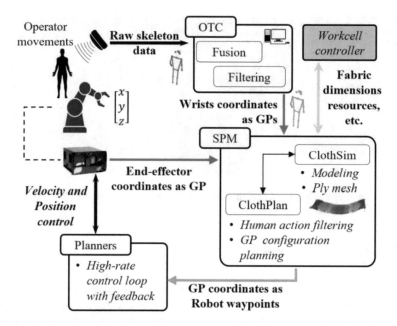

Fig. 3.7 Model-based fabric co-manipulation overall architecture [103], Reprinted with permission from IEEE

way [105]. A discrete artificial fish swarm algorithm for the assembly line balancing problem is presented in [106]. Its purpose is to minimize the costs and the number of stations in two-sided assembly lines.

George Michalos, Kousi, et al. (2018) presented a comprehensive and generalized approach for the implementation of a safe robotic system for HRC assembly, together with an orchestrator mechanism, based on the SoA concept [78]. The methodology suggested for the design and planning of collaborative applications, is presented in Fig. 3.8. It comprises seven steps, including the identification of the tasks that would benefit from HRC, given the experienced physical strain, the selection of the sequence of operations, aiming to end-up at the assignment of tasks to either humans or robots, based on suitability constraints; the approach was validated in a case study from the automotive industry. Subsequently, the layout of the respective workstation should be generated and then the definition of the collaborative workflow should take place.

This enables the breaking down of the workflow into distinct phases, depending on the mode of collaboration, to map them with the safety and the required by the safety standards specifications. Next, the system, including safety functions and interaction technologies is engineered i.e. the task execution, the safety behaviors, and the coordination of the resources are defined and the station controller module, together with the integration and communication architecture, are finely tuned. As the last step, the workflow and the safety concepts are integrated into the cell and the cell validation is performed.

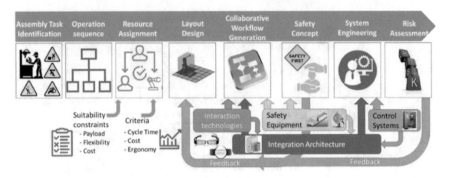

Fig. 3.8 Action workflow in an HRC paradigm for the automotive industry [78], Reprinted with permission from Pergamon

3.5 Machine Perception

Perception refers to the system's ability to become aware of its environment through the senses, where vision, touch, and hearing seem to be currently the most relevant areas in AI for manufacturing. Machine vision has enabled robots to autonomously perform operations such as navigation, handling, manipulation, and part processing, but also process monitoring, in-line inspections [107, 108], etc., and can thus play an important role in increasing the flexibility of manufacturing systems, as in [109]. Machine vision has also contributed to the safe and intuitive HRI, likewise audio and haptic signals processing. Audio processing allows the perception or generation of audio signals including not only speech but other sound material as well. Applications that have been enabled by audio processing include among others speech-based robot control, intuitive robot programming via voice commands, etc. Haptic signals can also be perceived by using force/torque, tactile, air-pressure sensors, or even proprioceptive sensors e.g., indicative examples for the aforementioned approaches are presented in the next paragraphs.

Visual servoing or vision-based robot control is the approach, where feedback from vision sensors is extracted to control a robot's motion. Visual servoing techniques are classified into image-based (control based on the deviation of the current and the target features on the image plane, no estimation of the target's pose is involved), the position/pose-based (the pose of the object is estimated having a camera as a reference and then a command is propagated to the robot controller) and hybrid approaches [110]. Furthermore, there are two configurations of the camera and end-effector; (a) the eye-in-hand configuration based on which the camera is mounted on the moving hand and observes the target's relative position e.g. [111], (b) the eye-to-hand where the camera is fixed in the surroundings of the robot and observes the target and the motion of the end-effector [112].

The prerequisite for the use of vision systems for the aforementioned functionalities to be enabled is camera calibration, which is the first step to ensure that the collection of three-dimensional data from the images is precise and hence the machine vision provides reliable results. In more detail, camera calibration techniques search for a set of image parameters, describing the mapping of 2D image reference coordinates to the 3D image reference coordinates. Neural Networks have been employed to compensate for lens distortion in camera calibration [113]. In addition, there are toolboxes and libraries ready to be used for camera calibration, namely the MATLAB toolbox [114] and the OpenCV (based on the Zhang method [115], which are also the most frequent ones used in research applications. Convolutional Neural Networks (CNN) seem to be the most popular technique for this purpose, as it enables calibration even in noise situations [116, 117].

There are various applications of vision systems that enable automated tasks. Sepp, Fuchs, and Hirzinger (2006) presented a hierarchical approach for object detection, initial-pose estimation, and real-time tracking based first on color distribution, shape, and texture information (histogram, particle filter, Mean-Shift) for part handling [118]. CNNs are frequently used to address recognition and detection tasks, as they have achieved in improving state-of-the-art accuracies [119–123]. Aivaliotis et al. applied a CNN-based approach for increasing the stability of robotized part manipulation, where errors in the manipulated part's position and orientation are identified after grasping [124]. The boosted random contextual semantic space [125], the first nearest neighbor classifier [126], the regional point descriptors [120], and the Deep Neural Networks have been employed for object detection and recognition, in robotic applications [127] as well as for industrial pallet classification and recognition of objects in noisy and cluttered scenes [128]. Some other strategies, which have also been proposed for object recognition and detection, are structural, probabilistic, graph-, feature- and physics-based algorithms, and hybrid strategies [129–133]. Fiducial markers and minimum complexity heuristics [134], and clustering-based algorithms [135]. A Deep Convolutional Activation Feature for Generic Visual Recognition (DeCAF) has been presented in tandem with a ready-to-use Python framework for easy network training [136].

Tasks including flexible parts have sparingly been studied for automation. This is due to challenges, such as the requirement for human-like sensitivity, as well as the numerous stochastic configurations that the parts can assume and high task complexity owing to the parts' nature. Additionally, lacking of a model for their deformation renders the manipulation of automation of flexible materials very challenging. Sardelis et al. (2021) proposed a model-free method, having considered the wide range of poses and deformations of deformable parts [137]. The proposed approach is low cost since it only requires 2D cameras; one to localize the grasping point (Fig. 3.9), and one to identify the cross-section. The perception algorithms are based on thresholding and depth of image analysis and they are easily adaptable to objects both similar in geometry and features.

The work in [137] presented a comprehensive solution for the automation of linear non-rigid components manipulation and assembly, which was the foundation for the later work of [112]. The manual operations were broken down into primitive actions, each of which had requirements in perceiving the surroundings, part position or type, etc. Therefore, perception systems for grasping point localization and cross-section recognition, as well as for the applied force and human–robot

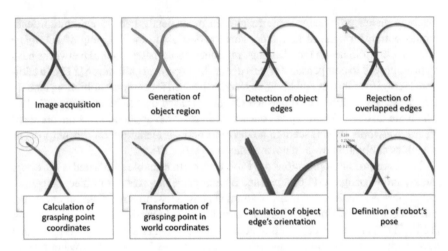

Fig. 3.9 Object detection and grasping point localization algorithm [137], Reprinted with permission from Elsevier BV

distance have been included, which in turn, provide feedback for robot control and affect the robot's behavior. The corresponding technological approach has addressed aspects, including safe flexible material supply, detection, and grasping. Figure 3.10 depicts the step-by-step operation that is enabled by an AI-based perception module), cross-section recognition (in the case of non-symmetrical parts), and assembly [112].

The completion of manufacturing tasks is usually followed by inspections and quality checks to prevent the propagation of errors to the imminent production steps. Neogi, Mohanta, and Dutta (2014) reviewed the vision-based approaches for steel surface inspection by presenting several methods e.g. spatial domain based, wavelet-based, using fractal model, support vector machine, and unsupervised classifier [138]. Weimer, Scholz-Reiter, and Shpitalni (2016) researched the automation of feature

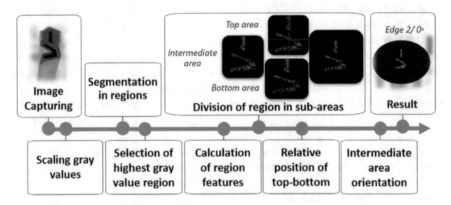

Fig. 3.10 Machine vision for flexible part cross-section recognition [112], Reprinted with permission from Pergamon

extraction for industrial inspections by proposing the use of CNN [139]. Artificial Neural Networks, in combination with image fusion, have been utilized for the detection of flaws in machined parts [140].

Another type of part that presents a challenge to perception systems is large parts, as in the majority of the scenarios, these parts cannot be portrayed in a single sensor frame. Prezas et al. (2022) worked on a multi-purpose perception system for dispensing applications in large linear parts where data for the process inspection are collected in-process to minimize cycle times [111]. The presented perception system has a threefold scope: (a) increasing the system's flexibility, (b) preventing process defects, and (c) providing a final decision on whether the process is successful or not (Fig. 3.11). In more detail, the first target is achieved by recognizing the part's type (deep convolutional neural networks) and position (color image processing) to trigger the correct robot processing routine, as well as to transform the offline routine into the actual part's position. Additionally, the same camera is used for the in-process detection of dispensed material discontinuities (HSV color model in tandem with a contour-based algorithm), and in-process detection of deviations (based on color image processing). Finally, the data that have been collected, while the process was running, are used for post-process quality control (DBSCAN clustering and convex hull calculations). This approach has been tested for a case study, inspired by the bus & coach sector.

The discussed approaches seem to provide promising results for enhancing robot perception however they usually require large volumes of annotated datasets to train the machine learning methods which is an expensive, susceptible to errors, and time-intensive process to a degree proportional to the complexity and the dynamic character of the operating environment. Manettas, Nikolakis, and Alexopoulos (2021) suggested the use of synthetic datasets as a solution for the aforementioned challenges, which are expected to accelerate the training process [141]. The presented work focuses on a framework for the generation of datasets through a chain of simulation tools, which generate several states of the parts of interest e.g. rotation in different rotation axes to be recognized by a computer-vision system. The authors tested their work for several CNN models, and they concluded that CNNs trained on synthetically generated datasets may have acceptable performance when used for supporting tasks in manufacturing.

The literature includes several works on speech-based HRI that are inspired by communication in human teams, where information transfer is achieved mostly by speech, and aims at increasing intuitiveness. This way of communication is mainly addressed to applications where humans and robots (usually cobots) can share a common workspace at the same time, and communication is deemed a key factor for the operator's safety and the acceptance of the robot. The core functionality of a speech recognition system is Natural Language Processing (NLP), however supporting modules and functions such as the recording device which includes wake-word detection and end of utterance detection, dialog management, and filtering of background noises, are necessary to implement a speech interface. Speech-based user interfaces have the advantage that they do not need the worker to have eye contact with the communication partner, as well as they can use their hands for other tasks [142].

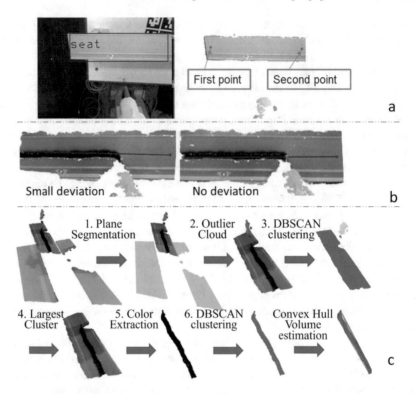

Fig. 3.11 a Profile type recognition (left) and part localization (right), **b** detection of path deviations, **c** Post-process quality control—point cloud manipulation steps [111], Reprinted with permission from Elsevier BV

The conversion of human instructions into robot actions requires speech recognition and speech processing, where usually the AI contribution is more intense, whereas the conversion process can be complex and computing power demanding. Thus, outsourcing speech processing to cloud-based speech recognition systems including Amazon Alexa [143] Google Dialogflow [144] Microsoft LUIS [145] is a common practice. Offline solutions such as Mozilla DeepSpeech or Rasa can be used instead when secure handling of the user data is prioritized.

A critical challenge is that the meaning of words frequently depends on the specific context (i.e. grounding is needed). Wölfel and Henrich (2020) dealt with the grounding problem by working on fuzzy logic for the mapping of uncertain instructions to control an industrial robot [146]. The performance of Speech Recognition-Systems can be assessed based on the delay under packet loss and accuracy [147]. Evaluated via the word error rate (comparison of the speech recognition results with a correct transcript of the utterance) and F1-score (true positives, false positives, true negatives, and false negatives for intent detection) Almansor and Hussain (2020) cate-gorized automated conversation systems into non-task-oriented, which allow users to participate in different domains but does not provide help into completing any task, and task-oriented, which are designed based on rules to help users achieve their goal or complete tasks [148]. Sotiris Makris et al. (2014) worked on task-oriented programming i.e. programming based on a library of predefined lower-level building blocks, for dual-arm robots, which they combined with gestures, voice commands, and graphical user interfaces for easier and modular robot programming [149]. Deuer-lein et al. (2020) designed and implemented a task-oriented software interface for HRC which recognized voice commands via cloud-based speech processing and subsequently converted them into machine-readable code [150]. Additionally, it was observed that for safe intent recognition when using a setup with a smart speaker, the sound ratio between the human utterance and the background noise should not exceed -8.75 dB.

In the paradigm of hybrid workstations, safe human–robot collaboration has been ensured through appropriate monitoring systems, tracking the human position or detecting collisions and triggering appropriate control strategies. AI has promoted the accuracy of the alarm generation and the smart adjustment of the robot's behavior for collision prevention or impact minimization. The calculation of the minimum distance on the fly between a human and a robot is usually needed to be used as a basis for the robot's behavior adjustments (Speed and Separation Monitoring, ISO/TS) [78]. Depth-based image processing and filtering have been proposed for collision avoidance [151], image processing using input from a stereo vision system and HSV (Hue, Saturation, Value) color space for detecting the humans together with 3D models of the robot and a rule-based system were presented to address false alarms by [152], 3D image analysis with the help of a physical model for the human's skin, along with motion analysis [153]. The potential fields method can enable the adjustment of the off-line planned robot paths based on field data from laser scanners, IMUs mounted on the operator, and the QR factorization method to compute the minimum distance between capsules representing humans and robots [154]. Nikolakis et al. (2019) presented a robot control module based on Cyber-Physical Systems together with depth data and heartbeat signals to dynamically control the robot's behavior by modifying its trajectory and speed in relation to human proximity [155].

Another approach taking advantage of the virtual space was used to enable the adaptation of the robots' behavior for flexible and reconfigurable systems [156]. This work is based on the Digital Twin concept and has been suggested in the context of hybrid production, where mobile dual arm workers autonomously navigate inside the shop floor undertaking multiple assembly operations, such as screwing, etc. while acting as assistants to human operators. This assembly paradigm requires high autonomy and flexible behavior from the robot side, which in turn necessitates orga-nizing efficiently all production entities and reason over the perceived environment using real-time data from the shop floor. To this end, a Digital Twin model is able to virtually represent in real-time the shop floor status and enable: (a) simplified control integration and sensor data sharing executed by the Execution Coordinator module which distributes the acquired data and tasks to all relevant resources such as mobile

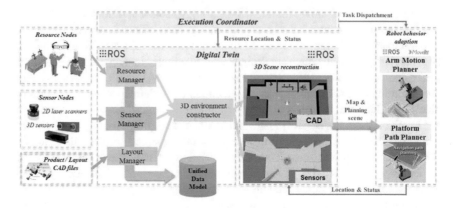

Fig. 3.12 Digital Twin for real-time robot adaptation [156], Reprinted with permission from Elsevier

robots, humans via HMIs, etc., (b) virtual representation of the shop floor continuously updated through a network of services using resource-related information by the Resource Manager, sing multiple sensor data combinations (Sensor Manager) and CAD models (Layout Manager), (c) generic unified semantic data modeling for semantically representing the geometrical as well as the workload state, (d) real-time robot behavior adaption by integrating standard robotic manipulator motion planners and mobile platform navigations planners to the Digital Twin for reasoning over the shop floor condition and producing safe and collision free paths during the different assembly tasks (Fig. 3.12).

Proprioceptive robot sensors have been frequently used together for the detection of collisions, via signal processing and dynamic thresholding-based approaches [157]. Kokkalis et al. (2018) also discussed an approach, based on proprioceptive robot sensors, to limit the forces applied by an industrial robot manipulator during contact [158]. The approach is based on the thresholding of the difference between the estimation of the current and the torque required by each joint for a given trajectory, as well as the actual current provided by the robot controller. The estimation is achieved via a time-invariant dynamic model, in combination with Artificial Neural Networks, whereas stop commands are issued in case the actual currents exceed the user-defined threshold (Fig. 3.13). This approach is implemented in the case of a low payload industrial robot and the results of the experiments show that the approach can provide good collision detection and force limiting results. Time-invariant dynamic models and supervised feedforward input-delay neural networks have also been suggested for similar purposes in [159].

Moreover, multi-sensor systems, including vision sensors, force sensors, and infrared proximity sensors, along with voice systems have been proposed to provide the operator with feedback on HRC [160]. Papanastasiou et al. (2019) presented a hybrid assembly workstation where manual guidance, air pressure contact sensors, and vision systems were combined for seamless human–robot collaboration [161].

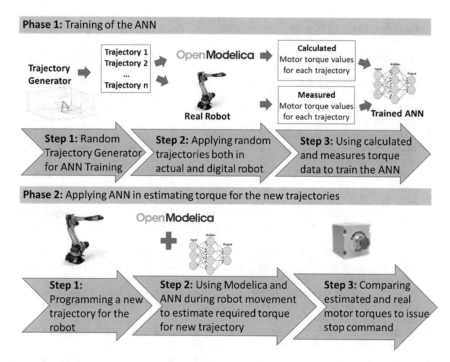

Fig. 3.13 Method overview for implementing power and force [158], Reprinted with permission from Elsevier BV

A force/torque sensor and impedance control were used for quick refinements of the robot's planned trajectories to the ongoing safety and process requirements. A shape detection and color segmentation-based vision system were used for the detection and feature tracking regardless of the accuracy in the initial positioning of the part to increase the accuracy of the robotized process (sealant application). Artificial Neural Networks were used in a thresholding-based implementation of the Power and Force limiting safety method (ISO/TS 15066), where motors' current and joints' position signals to execute a given trajectory are estimated via the simulation of a time-invariant dynamic model and are then compared to the actual current as provided by the robot controller.

References

1. Chryssolouris, G.: Manufacturing Systems: Theory and Practice. Springer (2006)
2. Lange, K.: Handbook of Metal Forming (1985)
3. Weck, M.: Handbook of Machine Tools. Metrological Analysis and Performance Test, vol. 4 (1984)
4. Romeo, L., Loncarski, J., Paolanti, M., Bocchini, G., Mancini, A., Frontoni, E.: Machine learning-based design support system for the prediction of heterogeneous machine parameters

in industry 4.0. Exp. Syst. Appl. **140**, 112869 (2020). https://doi.org/10.1016/J.ESWA.2019.
112869

5. Krings, A., Cossale, M., Tenconi, A., Soulard, J., Cavagnino, A., Boglietti, A.: Magnetic
   materials used in electrical machines: a comparison and selection guide for early machine
   design. IEEE Ind. Appl. Mag. **23**, 21–28 (2017). https://doi.org/10.1109/MIAS.2016.2600721

6. You, Y., Chen, C., Hu, F., Liu, Y., Ji, Z.: Advances of Digital Twins for predictive maintenance. Procedia Comput. Sci. **200**, 1471–1480 (2022). https://doi.org/10.1016/J.PROCS.
   2022.01.348

7. Schütze, A., Helwig, N., Schneider, T.: Sensors 4.0-smart sensors and measurement technology enable Industry 4.0. J. Sens. Sens. Syst. **7**, 359–371 (2018). https://doi.org/10.5194/
   jsss-7-359-2018

8. Dwivedy, S.K., Eberhard, P.: Dynamic analysis of flexible manipulators, a literature review
   (2006)

9. Moberg, S.: Modeling and Control of Flexible Manipulators (2010)

10. Health and Safety Executive: Selection and conformity of work equipment—Work equip-
    ment and machinery. https://www.hse.gov.uk/work-equipment-machinery/selection-confor
    mity.htm

11. Chatterjee, P., Athawale, V.M., Chakraborty, S.: Selection of industrial robots using compro-
    mise ranking and outranking methods. Robot. Comput. Integr. Manuf. **26**, 483–489 (2010).
    https://doi.org/10.1016/j.rcim.2010.03.007

12. Parkan, C., Wu, M.L.: Decision-making and performance measurement models with appli-
    cations to robot selection. Comput. Ind. Eng. **36**, 503–523 (1999). https://doi.org/10.1016/
    s0360-8352(99)00146-1

13. Braglia, M., Petroni, A.: Evaluating and selecting investments in industrial robots. Int. J. Prod.
    Res. **37**, 4157–4178 (1999). https://doi.org/10.1080/002075499189718

14. Chang, G.A., Sims, J.P.: A case-based reasoning approach to robot selection. In: American
    Society of Mechanical Engineers, Manufacturing Engineering Division, MED, pp. 943–951.
    American Society of Mechanical Engineers Digital Collection (2005)

15. Bhangale, P.P., Agrawal, V.P., Saha, S.K.: Attribute based specification, comparison and
    selection of a robot. In: Mechanism and Machine Theory, pp. 1345–1366. Pergamon (2004)

16. Rao, R.V., Padmanabhan, K.K.: Selection, identification and comparison of industrial robots
    using digraph and matrix methods. Robot. Comput. Integr. Manuf. **22**, 373–383 (2006). https://
    doi.org/10.1016/j.rcim.2005.08.003

17. Karsak, E.E.: Robot selection using an integrated approach based on quality function deploy-
    ment and fuzzy regression. Int. J. Prod. Res. **46**, 723–738 (2008). https://doi.org/10.1080/002
    07540600919571

18. Maldonado, A., Sánchez, J., Noriega, S., Díaz, J.J., García, J.L., Vidal, L.: A hierarchical fuzzy
    axiomatic design survey for ergonomic compatibility evaluation of advanced manufacturing
    technology–AMT. In: Proceedings of the 21st Annual International Occupational Ergonomics
    and Safety Conference, pp. 270–277 (2009)

19. Bahadir, M.C., Satoglu, S.I.: A novel robot arm selection methodology based on axiomatic
    design principles. Int. J. Adv. Manuf. Technol. **71**, 2043–2057 (2014). https://doi.org/10.1007/
    s00170-014-5620-2

20. Maldonado, A., García, J.L., Alvarado, A., Balderrama, C.O.: A hierarchical fuzzy axiomatic
    design methodology for ergonomic compatibility evaluation of advanced manufacturing tech-
    nology. Int. J. Adv. Manuf. Technol. **66**, 171–186 (2013). https://doi.org/10.1007/s00170-012-
    4316-8

21. Aly, M.F., Abbas, A.T., Megahed, S.M.: Robot workspace estimation and base placement
    optimisation techniques for the conversion of conventional work cells into autonomous flexible
    manufacturing systems. Int. J. Comput. Integr. Manuf. **23**, 1133–1148 (2010). https://doi.org/
    10.1080/0951192X.2010.528033

22. Michalos, G., Makris, S., Tsarouchi, P., Guasch, T., Kontovrakis, D., Chryssolouris, G.: Design
    considerations for safe human-robot collaborative workplaces. In: Procedia CIRP, pp. 248–
    253. Elsevier B.V. (2015)

23. Mgbemena, C.E., Tiwari, A., Xu, Y., Prabhu, V., Hutabarat, W.: Ergonomic evaluation on the manufacturing shop floor: a review of hardware and software technologies (2020)
24. Michalos, G., Karvouniari, A., Dimitropoulos, N., Togias, T., Makris, S.: Workplace analysis and design using virtual reality techniques. CIRP Ann. **67**, 141–144 (2018). https://doi.org/10.1016/j.cirp.2018.04.120
25. Savino, M.M., Battini, D., Riccio, C.: Visual management and artificial intelligence integrated in a new fuzzy-based full body postural assessment. Comput. Ind. Eng. **111**, 596–608 (2017). https://doi.org/10.1016/J.CIE.2017.06.011
26. Shirzadeh, M., Asl, H.J., Amirkhani, A., Jalali, A.A.: Vision-based control of a quadrotor utilizing artificial neural networks for tracking of moving targets. Eng. Appl. Artif. Intell. **58**, 34–48 (2017). https://doi.org/10.1016/J.ENGAPPAI.2016.10.016
27. Triboan, D., Chen, L., Chen, F., Wang, Z.: Semantic segmentation of real-time sensor data stream for complex activity recognition. Pers. Ubiquitous Comput. **21**, 411–425 (2017). https://doi.org/10.1007/S00779-017-1005-5/TABLES/4
28. Riboni, D., Murtas, M.: Sensor-based activity recognition: one picture is worth a thousand words. Futur. Gener. Comput. Syst. **101**, 709–722 (2019). https://doi.org/10.1016/J.FUTURE.2019.07.020
29. Heragu, S.S., Kusiak, A.: Machine layout problem in flexible manufacturing systems. **36**, 258–268 (1988). https://doi.org/10.1287/OPRE.36.2.258
30. Tubaileh, A.S.: Layout of robot cells based on kinematic constraints. **28**, 1142–1154 (2014). https://doi.org/10.1080/0951192X.2014.961552
31. Tsarouchi, P., Michalos, G., Makris, S., Athanasatos, T., Dimoulas, K., Chryssolouris, G.: On a human–robot workplace design and task allocation system. Int. J. Comput. Integr. Manuf. **30**, 1272–1279 (2017). https://doi.org/10.1080/0951192X.2017.1307524
32. Tao, L., Liu, Z.: Optimization on multi-robot workcell layout in vertical plane. In: 2011 IEEE International Conference on Information and Automation, ICIA 2011, pp. 744–749 (2011)
33. Kardos, C., Kovács, A., Váncza, J.: Towards feature-based human-robot assembly process planning. In: Procedia CIRP, pp. 516–521. Elsevier B.V. (2016)
34. Tsarouchi, P., Makris, S., Chryssolouris, G.: Human–robot interaction review and challenges on task planning and programming. Int. J. Comput. Integr. Manuf. **29**, 916–931 (2016). https://doi.org/10.1080/0951192X.2015.1130251
35. Agostini, A., Torras, C., Wörgötter, F.: Integrating task planning and interactive learning for robots to work in human environments. In: IJCAI International Joint Conference on Artificial Intelligence, pp. 2386–2391 (2011)
36. Takata, S., Hirano, T.: Human and robot allocation method for hybrid assembly systems. CIRP Ann. Manuf. Technol. **60**, 9–12 (2011). https://doi.org/10.1016/j.cirp.2011.03.128
37. Malvankar-Mehta, M.S., Mehta, S.S.: Optimal task allocation in multi-human multi-robot interaction. Optim. Lett. **9**, 1787–1803 (2015). https://doi.org/10.1007/s11590-015-0890-7
38. Tsarouchi, P., Matthaiakis, A.-S., Makris, S., Chryssolouris, G.: On a human-robot collaboration in an assembly cell. Int. J. Comput. Integr. Manuf. **30**, 580–589 (2017). https://doi.org/10.1080/0951192X.2016.1187297
39. Nikolakis, N., Kousi, N., Michalos, G., Makris, S.: Dynamic scheduling of shared human-robot manufacturing operations. Proc. CIRP **72**, 9–14 (2018). https://doi.org/10.1016/j.procir.2018.04.007
40. Alami, R., Chatila, R., Fleury, S., Ghallab, M., Ingrand, F.: An architecture for autonomy. Int. J. Robot. Res. Spec. Issue Integr. Arch. Robot Control Program. **17**, 315–337 (1998).https://doi.org/10.2307/j.ctv36zr6w.7
41. Alili, S., Warnier, M., Ali, M., Alami, R.: Planning and plan-execution for human-robot cooperative task achievement decisional architecture for human robot. In: 4th Workshop on Planning and Plan Execution for Real-World Systems Principles and Practices for Planning in Execution (2009)
42. Chuan Tan, J.T., Duan, F., Kato, R., Arai, T.: Collaboration planning by task analysis in human-robot collaborative manufacturing system. In: Advances in Robot Manipulators (2010)

43. Michalos, G., Spiliotopoulos, J., Makris, S., Chryssolouris, G.: A method for planning human robot shared tasks. CIRP J. Manuf. Sci. Technol. **22**, 76–90 (2018). https://doi.org/10.1016/j.cirpj.2018.05.003

44. Garrett, C.R., Chitnis, R., Holladay, R., Kim, B., Silver, T., Kaelbling, L.P., Lozano-Perez, T.: Integrated Task and Motion Planning (2021)

45. Veeramani, S., Muthuswamy, S., Sagar, K., Zoppi, M.: Multi-head path planning of SwarmItFIX agents: a Markov decision process approach. Mech. Mach. Sci. **73**, 2237–2247 (2019). https://doi.org/10.1007/978-3-030-20131-9_221

46. Ahmad, R., Plapper, P.: Safe and automated tool-path generation for multi-axis production machines. In: ASME International Mechanical Engineering Congress and Exposition, Proceedings (IMECE), vol. 2B, pp. 1–7 (2014). https://doi.org/10.1115/IMECE2014-36742

47. Fichtner, D., Nestler, A., Dang, T.N., Schulze, A., Carlsen, U., Schreiber, S., Lee, S.W.: Use of agents and neural networks for acquisition and preparation of distributed NC information to support NC planning. **19**, 581–592 (2007). https://doi.org/10.1080/09511920600651956

48. Eberspächer, P., Schraml, P., Schlechtendahl, J., Verl, A., Abele, E.: A model- and signal-based power consumption monitoring concept for energetic optimization of machine tools. Proc. CIRP **15**, 44–49 (2014). https://doi.org/10.1016/J.PROCIR.2014.06.020

49. Cai, Y., Shi, X., Shao, H., Wang, R., Liao, S.: Energy efficiency state identification in milling processes based on information reasoning and Hidden Markov Model. J. Clean. Prod. **193**, 397–413 (2018). https://doi.org/10.1016/J.JCLEPRO.2018.04.265

50. Thomas, U., Wahl, F.M.: A system for automatic planning, evaluation and execution of assembly sequences for industrial robots. In: IEEE International Conference on Intelligent Robots and Systems, pp. 1458–1464 (2001)

51. Lepuschitz, W., Zoitl, A., Merdan, M.: Ontology-driven automated software configuration for manufacturing system components. In: Conference Proceedings—IEEE International Conference on Systems, Man and Cybernetics, pp. 427–433 (2011)

52. Stenmark, M., Malec, J.: Knowledge-based instruction of manipulation tasks for industrial robotics. Robot. Comput. Integr. Manuf. **33**, 56–67 (2015). https://doi.org/10.1016/j.rcim.2014.07.004

53. Fox, M., Long, D.: PDDL2.1: An extension to PDDL for expressing temporal planning domains. journal of artificial intelligence research. **20**, 61–124 (2003). https://doi.org/10.1613/JAIR.1129

54. Kaltsoukalas, K., Makris, S., Chryssolouris, G.: On generating the motion of industrial robot manipulators. Robot. Comput. Integr. Manuf. **32**, 65–71 (2015). https://doi.org/10.1016/j.rcim.2014.10.002

55. Zöllner, R., Asfour, T., Dillmann, R.: Programming by demonstration: dual-arm manipulation tasks for humanoid robots. In: 2004 IEEE/RSJ International Conference on Intelligent Robots and Systems (IROS), pp. 479–484 (2004)

56. Tsianos, K.I., Sucan, I.A., Kavraki, L.E.: Sampling-based robot motion planning: towards realistic applications. Comput. Sci. Rev. **1**, 2–11 (2007). https://doi.org/10.1016/j.cosrev.2007.08.002

57. Kavraki, L.E., Švestka, P., Latombe, J.C., Overmars, M.H.: Probabilistic roadmaps for path planning in high-dimensional configuration spaces. IEEE Trans. Robot. Autom. **12**, 566–580 (1996). https://doi.org/10.1109/70.508439

58. Bayazit, O.B., Lien, J.M., Amato, N.M.: Probabilistic roadmap motion planning for deformable objects. In: Proceedings—IEEE International Conference on Robotics and Automation, pp. 2126–2133 (2002)

59. Koo, K., Jiang, X., Konno, A., Uchiyama, M.: Development of a wire harness assembly motion planner for redundant multiple manipulators. J. Robot. Mechatronics. **23**, 907–918 (2011). https://doi.org/10.20965/jrm.2011.p0907

60. Kohrt, C., Stamp, R., Pipe, A.G., Kiely, J., Schiedermeier, G.: An online robot trajectory planning and programming support system for industrial use. Robot. Comput. Integr. Manuf. **29**, 71–79 (2013). https://doi.org/10.1016/j.rcim.2012.07.010

61. Lavalle, S.M., Lavalle, S.M.: Rapidly-Exploring Random Trees: A New Tool for Path Planning (1998)
62. Ferguson, D., Kalra, N., Stentz, A.: Replanning with RRTs. In: Proc IEEE International Conference on Robotics and Automation 2006, pp. 1243–1248 (2006). https://doi.org/10.1109/ROBOT.2006.1641879
63. Chong, J.W.S., Ong, S.K., Nee, A.Y.C., Youcef-Youmi, K.: Robot programming using augmented reality: an interactive method for planning collision-free paths. Robot. Comput. Integr. Manuf. **25**, 689–701 (2009). https://doi.org/10.1016/j.rcim.2008.05.002
64. Wu, X.J., Tang, J., Li, Q., Heng, K.H.: Development of a configuration space motion planner for robot in dynamic environment. Robot. Comput. Integr. Manuf. **25**, 13–31 (2009). https://doi.org/10.1016/j.rcim.2007.04.004
65. Qureshi, A.H., Ayaz, Y.: Intelligent bidirectional rapidly-exploring random trees for optimal motion planning in complex cluttered environments. Robot. Auton. Syst. **68**, 1–11 (2015). https://doi.org/10.1016/J.ROBOT.2015.02.007
66. Glorieux, E., Franciosa, P., Ceglarek, D.: Quality and productivity driven trajectory optimisation for robotic handling of compliant sheet metal parts in multi-press stamping lines. Robot. Comput. Integr. Manuf. **56**, 264–275 (2019). https://doi.org/10.1016/J.RCIM.2018.10.004
67. Patle, B.K., Babu L, G., Pandey, A., Parhi, D.R.K., Jagadeesh, A.: A review: on path planning strategies for navigation of mobile robot (2019)
68. Nayab Zafar, M., Mohanta, J.C., Sanyal, A.: Design and implementation of an autonomous robot manipulator for pick & place planning. IOP Conf. Ser. Mater. Sci. Eng. **691**, 012008 (2019). https://doi.org/10.1088/1757-899X/691/1/012008
69. Zhu, J., Navarro, B., Passama, R., Fraisse, P., Crosnier, A., Cherubini, A.: Robotic manipulation planning for shaping deformable linear objects with environmental contacts. IEEE Robot. Autom. Lett. **5**, 16–23 (2020). https://doi.org/10.1109/LRA.2019.2944304
70. King, J.E., Cognetti, M., Srinivasa, S.S.: Rearrangement planning using object-centric and robot-centric action spaces. In: Proceedings of IEEE International Conference on Robotics and Automation, 2016-June, pp. 3940–3947 (2016). https://doi.org/10.1109/ICRA.2016.7487583
71. Peng, J.: Mobile robot path planning based on improved Q learning algorithm. Int. J. Multimed. Ubiquitous Eng. **10**, 285–294 (2015). https://doi.org/10.14257/ijmue.2015.10.7.30
72. Xu, F., Van Brussel, H., Nuttin, M., Moreas, R.: Concepts for dynamic obstacle avoidance and their extended application in underground navigation. Robot. Auton. Syst. **42**, 1–15 (2003). https://doi.org/10.1016/S0921-8890(02)00323-8
73. Zhang, Z., Yue, S., Zhang, G.: Fly visual system inspired artificial neural network for collision detection. Neurocomputing **153**, 221–234 (2015). https://doi.org/10.1016/j.neucom.2014.11.033
74. Hoshino, S., Maki, K.: Safe and efficient motion planning of multiple mobile robots based on artificial potential for human behavior and robot congestion. Adv. Robot. **29**, 1095–1109 (2015). https://doi.org/10.1080/01691864.2015.1033461
75. Zi, B., Lin, J., Qian, S.: Localization, obstacle avoidance planning and control of a cooperative cable parallel robot for multiple mobile cranes. Robot. Comput. Integr. Manuf. **34**, 105–123 (2015). https://doi.org/10.1016/j.rcim.2014.11.005
76. Das, P.K., Behera, H.S., Panigrahi, B.K.: A hybridization of an improved particle swarm optimization and gravitational search algorithm for multi-robot path planning. Swarm Evol. Comput. **28**, 14–28 (2016). https://doi.org/10.1016/J.SWEVO.2015.10.011
77. Makris, S., Kampourakis, E., Andronas, D.: On deformable object handling: model-based motion planning for human-robot co-manipulation. CIRP Ann. (2022). https://doi.org/10.1016/J.CIRP.2022.04.048
78. Michalos, G., Kousi, N., Karagiannis, P., Gkournelos, C., Dimoulas, K., Koukas, S., Mparis, K., Papavasileiou, A., Makris, S.: Seamless human robot collaborative assembly—an automotive case study. Mechatronics **55**, 194–211 (2018). https://doi.org/10.1016/j.mechatronics.2018.08.006

79. Viola, C. La, Orlandini, A., Umbrico, A., Cesta, A.: ROS-TiPlEx: a collaborative design tool for timeline-based planning & scheduling applications with ROS (2020)

80. Evangelou, G., Dimitropoulos, N., Michalos, G., Makris, S.: An approach for task and action planning in human-robot collaborative cells using AI. Proc. CIRP **97**, 476–481 (2020). https://doi.org/10.1016/j.procir.2020.08.006

81. Cesta, A., Orlandini, A., Umbrico, A.: Fostering robust human-robot collaboration through AI task planning. Proc. CIRP **72**, 1045–1050 (2018). https://doi.org/10.1016/J.PROCIR.2018.03.022

82. Farnworth, M.J., Barrett, L.A., Adams, N.J., Beausoleil, N.J., Weidgraaf, K., Hekman, M., Chambers, J.P., Thomas, D.G., Waran, N.K., Stafford, K.J.: Assessment of a carbon dioxide laser for the measurement of thermal nociceptive thresholds following intramuscular administration of analgesic drugs in pain-free female cats. Vet. Anaesth. Analg. **42**, 638–647 (2015). https://doi.org/10.1111/vaa.12245

83. Mayer, MPh., Odenthal, B., Faber, M., Winkelholz, C., Schlick, C.M.: Cognitive engineering of automated assembly processes. Hum. Factors Ergon. Manuf. Serv. Ind. **24**, 348–368 (2014). https://doi.org/10.1002/hfm.20390

84. Michalos, G., Karagiannis, P., Makris, S., Tokçalar, Ö., Chryssolouris, G.: Augmented reality (AR) applications for supporting human-robot interactive cooperation. Proc. CIRP **41**, 370–375 (2016). https://doi.org/10.1016/j.procir.2015.12.005

85. Liu, H., Wang, L.: An AR-based worker support system for human-robot collaboration. Procedia Manuf. **11**, 22–30 (2017). https://doi.org/10.1016/j.promfg.2017.07.124

86. Liu, H., Wang, L.: Gesture recognition for human-robot collaboration: a review. Int. J. Ind. Ergon. **68**, 355–367 (2018). https://doi.org/10.1016/j.ergon.2017.02.004

87. Silaghi, H., Rohde, U., Spoiala, V., Silaghi, A., Gergely, E., Nagy, Z.: Voice command of an industrial robot in a noisy environment. In: 2014 International Symposium on Fundamentals of Electrical Engineering, ISFEE 2014. Institute of Electrical and Electronics Engineers Inc. (2015)

88. Hogreve, S., Kaczmarek, S., Adam, J., Franz, L., Döllen, T., Paulus, H., Reinkemeyer, V., Tracht, K.: Controlling and assisting manual assembly processes by automated progress and gesture recognition. Appl. Mech. Mater. **840**, 50–57 (2016). https://doi.org/10.4028/www.scientific.net/amm.840.50

89. Kaczmarek, S., Hogreve, S., Tracht, K.: Progress monitoring and gesture control in manual assembly systems using 3D-image sensors. In: Procedia CIRP, pp. 1–6. Elsevier B.V. (2015)

90. Iba, S., Paredis, C.J.J., Khosla, P.K.: Intention aware interactive multi-modal robot programming. In: IEEE International Conference on Intelligent Robots and Systems, pp. 3479–3484 (2003)

91. Huber, A., Weiss, A.: Developing human-robot interaction for an industry 4.0 robot: how industry workers helped to improve remote-hri to physical-hri. In: ACM/IEEE International Conference on Human-Robot Interaction, pp. 137–138. IEEE Computer Society, New York, NY, USA (2017)

92. Ang, M.H., Lin, W., Lim, S.Y.: Walk-through programmed robot for welding in shipyards. Ind. Robot. **26**, 377–388 (1999). https://doi.org/10.1108/01439919910284000/FULL/PDF

93. Neto, P., Pires, J., Moreira, A.: High-level programming for industrial robotics: using gestures, speech and force control. Ind. Robot Int. J. **37**, 137–147 (2009). https://doi.org/10.1108/01439911011018911

94. Gkournelos, C., Karagiannis, P., Kousi, N., Michalos, G., Koukas, S., Makris, S.: Application of wearable devices for supporting operators in human-robot cooperative assembly tasks. In: Procedia CIRP, pp. 177–182. Elsevier B.V. (2018)

95. Dimitropoulos, N., Togias, T., Michalos, G., Makris, S.: Operator support in human-robot collaborative environments using AI enhanced wearable devices. In: Procedia CIRP, pp. 464–469. Elsevier (2020)

96. Apostolopoulos, G., Andronas, D., Fourtakas, N., Makris, S.: Operator training framework for hybrid environments: an augmented reality module using machine learning object recognition. Procedia CIRP **106**, 102–107 (2022). https://doi.org/10.1016/J.PROCIR.2022.02.162

97. Matthaiakis, S.A., Dimoulas, K., Athanasatos, A., Mparis, K., Dimitrakopoulos, G., Gkour-nelos, C., Papavasileiou, A., Fousekis, N., Papanastasiou, S., Michalos, G., Angione, G., Makris, S.: Flexible programming tool enabling synergy between human and robot. Procedia Manuf. **11**, 431–440 (2017). https://doi.org/10.1016/j.promfg.2017.07.131

98. Sangdani, M.H., Tavakolpour-Saleh, A.R., Lotfavar, A.: Genetic algorithm-based optimal computed torque control of a vision-based tracker robot: simulation and experiment. Eng. Appl. Artif. Intell. **67**, 24–38 (2018). https://doi.org/10.1016/J.ENGAPPAI.2017.09.014

99. Szczepanski, R., Tarczewski, T., Grzesiak, L.M.: Adaptive state feedback speed controller for PMSM based on Artificial Bee Colony algorithm. Appl. Softw. Comput. **83**, 105644 (2019). https://doi.org/10.1016/J.ASOC.2019.105644

100. Fister, D., Fister, I., Fister, I., Šafarič, R.: Parameter tuning of PID controller with reactive nature-inspired algorithms. Robot. Auton. Syst. **84**, 64–75 (2016). https://doi.org/10.1016/J.ROBOT.2016.07.005

101. Son, C.: Intelligent jamming region division with machine learning and fuzzy optimization for control of robot's part micro-manipulative task. Inf. Sci. (N Y) **256**, 211–224 (2014). https://doi.org/10.1016/J.INS.2013.08.008

102. Nagabandi, A., Konolige, K., Levine, S., Kumar, V., Brain, G.: Deep Dynamics Models for Learning Dexterous Manipulation (2020)

103. Andronas, D., Kampourakis, E., Bakopoulou, K., Gkournelos, C., Angelakis, P., Makris, S.: Model-based robot control for human-robot flexible material co-manipulation. In: 2021 26th IEEE International Conference on Emerging Technologies and Factory Automation (ETFA), pp. 1–8 (2021). https://doi.org/10.1109/ETFA45728.2021.9613235

104. Ding, K., Lei, J., Chan, F.T.S., Hui, J., Zhang, F., Wang, Y.: Hidden Markov model-based autonomous manufacturing task orchestration in smart shop floors. Robot. Comput. Integr. Manuf. **61**, 101845 (2020). https://doi.org/10.1016/J.RCIM.2019.101845

105. Joo, T., Seo, M., Shin, D.: An adaptive approach for determining batch sizes using the hidden Markov model. J. Intell. Manuf. **30**, 917–932 (2019). https://doi.org/10.1007/S10845-017-1297-3/FIGURES/8

106. Zhong, Y., Deng, Z., Xu, K.: An effective artificial fish swarm optimization algorithm for two-sided assembly line balancing problems. Comput. Ind. Eng. **138**, 106121 (2019). https://doi.org/10.1016/J.CIE.2019.106121

107. Ker, J., Kengskool, K.: An efficient method for inspecting machined parts by a fixtureless machine vision system. Society of Manufacturing Engineers (1990)

108. Sanz, J.L.C., Petković, D.: Machine vision algorithms for automated inspection of thin-film disk heads. IEEE Trans. Pattern Anal. Mach. Intell. **10**, 830–848 (1988). https://doi.org/10.1109/34.9106

109. Golnabi, H., Asadpour, A.: Design and application of industrial machine vision systems. Robot. Comput. Integr. Manuf. **23**, 630–637 (2007). https://doi.org/10.1016/j.rcim.2007.02.005

110. Torralba, A., Murphy, K.P., Freeman, W.T., Rubin, M.A.: Context-based vision system for place and object recognition. In: Proceedings of the IEEE International Conference on Computer Vision, vol. 1, pp. 273–280 (2003).https://doi.org/10.1109/iccv.2003.1238354

111. Prezas, L., Michalos, G., Arkouli, Z., Katsikarelis, A., Makris, S.: AI-enhanced vision system for dispensing process monitoring and quality control in manufacturing of large parts. Procedia CIRP **107**, 1275–1280 (2022). https://doi.org/10.1016/j.procir.2022.05.144

112. Andronas, D., Arkouli, Z., Zacharaki, N., Michalos, G., Sardelis, A., Papanikolopoulos, G., Makris, S.: On the perception and handling of deformable objects—a robotic cell for white goods industry. Robot. Comput. Integr. Manuf. **77**, 102358 (2022). https://doi.org/10.1016/j.rcim.2022.102358

113. Chung, B.M.: Neural-network model for compensation of lens distortion in camera calibration. Int. J. Precis. Eng. Manuf. **19**, 959–966 (2018). https://doi.org/10.1007/s12541-018-0113-0

114. Bouguet, J.: Camera calibration toolbox for Matlab. Computational Vision at the California Institute of Technology (2012)

115. Zhang, Z.: A flexible new technique for camera calibration. IEEE Trans. Pattern Anal. Mach. Intell. **22**, 1330–1334 (2000). https://doi.org/10.1109/34.888718
116. Raza, S.N., Raza Ur Rehman, H., Lee, S.G., Sang Choi, G.: Artificial intelligence based camera calibration. In: 2019 15th International Wireless Communications and Mobile Computing Conference, IWCMC 2019, pp. 1564–1569. Institute of Electrical and Electronics Engineers Inc. (2019)
117. Itu, R., Borza, D., Danescu, R.: Automatic extrinsic camera parameters calibration using convolutional neural networks. In: Proceedings—2017 IEEE 13th International Conference on Intelligent Computer Communication and Processing, ICCP 2017, pp. 273–278. Institute of Electrical and Electronics Engineers Inc. (2017)
118. Sepp, W., Fuchs, S., Hirzinger, G.: Hierarchical featureless tracking for position-based 6-DoF visual servoing. In: IEEE International Conference on Intelligent Robots and Systems, pp. 4310–4315 (2006). https://doi.org/10.1109/IROS.2006.281964
119. Liang, M., Hu, X.: Recurrent Convolutional Neural Network for Object Recognition (2015)
120. Frome, A., Huber, D., Kolluri, R., Bülow, T., Malik, J.: Recognizing objects in range data using regional point descriptors. Lecture Notes in Computer Science (including subseries Lecture Notes in Artificial Intelligence and Lecture Notes in Bioinformatics), vol. 3023, pp. 224–237 (2004).https://doi.org/10.1007/978-3-540-24672-5_18
121. Ciregan, D., Meier, U., Schmidhuber, J.: Multi-column deep neural networks for image classification. In: Proceedings of the IEEE Computer Society Conference on Computer Vision and Pattern Recognition, pp. 3642–3649 (2012)
122. He, K., Sun, J.: Convolutional Neural Networks at Constrained Time Cost (2015)
123. Andrianakos, G., Dimitropoulos, N., Michalos, G., Makris, S.: An approach for monitoring the execution of human based assembly operations using machine learning. In: Procedia CIRP, pp. 198–203. Elsevier B.V. (2020)
124. Aivaliotis, P., Zampetis, A., Michalos, G., Makris, S.: A machine learning approach for visual recognition of complex parts in robotic manipulation. Procedia Manuf. **11**, 423–430 (2017). https://doi.org/10.1016/j.promfg.2017.07.130
125. Zhang, C., Xue, Z., Zhu, X., Wang, H., Huang, Q., Tian, Q.: Boosted random contextual semantic space based representation for visual recognition. Inf. Sci. (N Y) **369**, 160–170 (2016). https://doi.org/10.1016/j.ins.2016.06.029
126. Bai, X., Yang, X., Latecki, L.J.: Detection and recognition of contour parts based on shape similarity. Pattern Recognit. **41**, 2189–2199 (2008). https://doi.org/10.1016/j.patcog.2007.12.016
127. Lu, K., An, X., Li, J., He, H.: Efficient deep network for vision-based object detection in robotic applications. Neurocomputing **245**, 31–45 (2017). https://doi.org/10.1016/j.neucom.2017.03.050
128. Rendall, R., Castillo, I., Lu, B., Colegrove, B., Broadway, M., Chiang, L.H., Reis, M.S.: Image-based manufacturing analytics: improving the accuracy of an industrial pellet classification system using deep neural networks. Chemom. Intell. Lab. Syst. **180**, 26–35 (2018). https://doi.org/10.1016/j.chemolab.2018.07.001
129. Stefańczyk, M., Pietruch, R.: Hypothesis generation in generic, model-based object recognition system. In: Advances in Intelligent Systems and Computing, pp. 717–727. Springer (2016)
130. Lutz, M., Stampfer, D., Schlegel, C.: Probabilistic object recognition and pose estimation by fusing multiple algorithms. In: Proceedings—IEEE International Conference on Robotics and Automation, pp. 4244–4249 (2013)
131. Wang, X., Lin, L., Huang, L., Yan, S.: Incorporating structural alternatives and sharing into hierarchy for multiclass object recognition and detection. 3334–3341 (2013). https://doi.org/10.1109/CVPR.2013.428
132. Karambakhsh, A., Sheng, B., Li, P., Yang, P., Jung, Y., Feng, D.D.: VoxRec: hybrid convolutional neural network for active 3D object recognition. IEEE Access. **8**, 70969–70980 (2020). https://doi.org/10.1109/ACCESS.2020.2987177

133. Li, H., Lin, J.C.: Using fuzzy logic to detect dimple defects of polished wafer surfaces. IEEE Trans. Ind. Appl. **30**, 317–323 (1994). https://doi.org/10.1109/28.287528

134. Wang, J., Olson, E.: AprilTag 2: Efficient and robust fiducial detection. In: IEEE International Conference on Intelligent Robots and Systems, pp. 4193–4198. Institute of Electrical and Electronics Engineers Inc. (2016)

135. Hillenbrand, U.: Pose Clustering from Stereo Data (2008)

136. Donahue, J., Jia, Y., Vinyals, O., Hoffman, J., Zhang, N., Tzeng, E., Darrell, T.: DeCAF: A Deep Convolutional Activation Feature for Generic Visual Recognition (2014)

137. Sardelis, A., Zacharaki, N.-C., Arkouli, Z., Andronas, D., Michalos, G., Makris, S., Papanikolopoulos, G.: 2-Stage vision system for robotic handling of flexible objects. Proc. CIRP **97**, 491–496 (2021). https://doi.org/10.1016/j.procir.2020.07.008

138. Neogi, N., Mohanta, D.K., Dutta, P.K.: Review of vision-based steel surface inspection systems. J. Comput. High Educ. **2014**, 1–19 (2014). https://doi.org/10.1186/1687-5281-2014-50/TABLES/5

139. Weimer, D., Scholz-Reiter, B., Shpitalni, M.: Design of deep convolutional neural network architectures for automated feature extraction in industrial inspection. CIRP Ann. **65**, 417–420 (2016). https://doi.org/10.1016/J.CIRP.2016.04.072

140. Satorres Martínez, S., Ortega Vázquez, C., Gámez García, J., Gómez Ortega, J.: Quality inspection of machined metal parts using an image fusion technique. Measurement **111**, 374–383 (2017). https://doi.org/10.1016/J.MEASUREMENT.2017.08.002

141. Manettas, C., Nikolakis, N., Alexopoulos, K.: Synthetic datasets for Deep Learning in computer-vision assisted tasks in manufacturing. Procedia CIRP **103**, 237–242 (2021). https://doi.org/10.1016/J.PROCIR.2021.10.038

142. Makris, S.: Cooperating Robots for Flexible Manufacturing (2021)

143. Amazon: Alexa Skills Kit. https://developer.amazon.com/en-GB/docs/alexa/ask-overviews/what-is-the-alexa-skills-kit.html

144. Google: Dialogflow API. https://cloud.google.com/dialogflow/docs/

145. Microsoft: Language Understanding (LUIS) Documentation: learn how language understanding enables your applications to understand what a person wants in their own words. https://docs.microsoft.com/en-gb/azure/cognitive-services/luis/

146. Wölfel, K., Henrich, D.: Grounding of uncertain force parameters in spoken robot commands. Adv. Intell. Syst. Comput. **980**, 194–201 (2020). https://doi.org/10.1007/978-3-030-19648-6_23

147. Assefi, M., Wittie, M., Knight, A.: Impact of network performance on cloud speech recognition. In: Proceedings—International Conference on Computer Communications and Networks, ICCCN. 2015-October, (2015). https://doi.org/10.1109/ICCCN.2015.7288417

148. Almansor, E.H., Hussain, F.K.: Survey on intelligent chatbots: state-of-the-art and future research directions. Adv. Intell. Syst. Comput. **993**, 534–543 (2020). https://doi.org/10.1007/978-3-030-22354-0_47

149. Makris, S., Tsarouchi, P., Surdilovic, D., Krüger, J.: Intuitive dual arm robot programming for assembly operations. CIRP Ann. Manuf. Technol. **63**, 13–16 (2014). https://doi.org/10.1016/j.cirp.2014.03.017

150. Deuerlein, C., Langer, M., Seßner, J., Heß, P., Franke, J.: Human-robot-interaction using cloud-based speech recognition systems. Procedia CIRP **97**, 130–135 (2020). https://doi.org/10.1016/j.procir.2020.05.214

151. Schmidt, B., Wang, L.: Contact-less and programming-less human-robot collaboration. In: Procedia CIRP, pp. 545–550. Elsevier B.V. (2013)

152. Grabowski, A., Kosiński, R., Dźwiarek, M.: Vision based safety system for human and robot arm detection. In: IFAC Proceedings Volumes (IFAC-PapersOnline), pp. 68–72. IFAC Secretariat (2009)

153. Krüger, J., Nickolay, B., Heyer, P., Seliger, G.: Image based 3D surveillance for flexible man-robot-cooperation. CIRP Ann. Manuf. Technol. **54**, 19–22 (2005). https://doi.org/10.1016/S0007-8506(07)60040-7

154. Safeea, M., Neto, P.: Minimum distance calculation using laser scanner and IMUs for safe human-robot interaction. Robot. Comput. Integr. Manuf. **58**, 33–42 (2019). https://doi.org/10.1016/j.rcim.2019.01.008

155. Nikolakis, N., Maratos, V., Makris, S.: A cyber physical system (CPS) approach for safe human-robot collaboration in a shared workplace. Robot. Comput. Integr. Manuf. **56**, 233–243 (2019). https://doi.org/10.1016/j.rcim.2018.10.003

156. Kousi, N., Gkournelos, C., Aivaliotis, S., Giannoulis, C., Michalos, G., Makris, S.: Digital twin for adaptation of robots' behavior in flexible robotic assembly lines. In: Procedia Manufacturing, pp. 121–126. Elsevier B.V. (2019)

157. De Luca, A., Albu-Schäffer, A., Haddadin, S., Hirzinger, G.: Collision detection and safe reaction with the DLR-III lightweight manipulator arm. In: IEEE International Conference on Intelligent Robots and Systems, pp. 1623–1630 (2006)

158. Kokkalis, K., Michalos, G., Aivaliotis, P., Makris, S.: An approach for implementing power and force limiting in sensorless industrial robots. Procedia CIRP **76**, 138–143 (2018). https://doi.org/10.1016/j.procir.2018.01.028

159. Aivaliotis, P., Aivaliotis, S., Gkournelos, C., Kokkalis, K., Michalos, G., Makris, S.: Power and force limiting on industrial robots for human-robot collaboration. Robot. Comput. Integr. Manuf. **59**, 346–360 (2019). https://doi.org/10.1016/j.rcim.2019.05.001

160. Bdiwi, M.: Integrated sensors system for human safety during cooperating with industrial robots for handing-over and assembling tasks. Procedia CIRP **23**, 65–70 (2014). https://doi.org/10.1016/j.procir.2014.10.099

161. Papanastasiou, S., Kousi, N., Karagiannis, P., Gkournelos, C., Papavasileiou, A., Dimoulas, K., Baris, K., Koukas, S., Michalos, G., Makris, S.: Towards seamless human robot collaboration: integrating multimodal interaction. Int. J. Adv. Manuf. Technol. **105**, 3881–3897 (2019). https://doi.org/10.1007/s00170-019-03790-3

Chapter 4
Artificial Intelligence in Manufacturing Systems

Abstract A manufacturing system is assumed to be comprising a combination of machines, cells, intra-logistics devices and other peripheral devices, used on the factory floor as well as on logistics. This chapter focuses on AI at the manufacturing system level. At system level, the volume and a variety of relevant data increase, activities are characterized by a higher degree of uncertainty and stochasticity, with several interdependencies among the parameters in non-linear relations. All these inherited attributes make transparency, predictability and adaptability more challenging tasks for AI. More specifically, the chapter examines (i) AI for the design of a manufacturing system that relates the design to process and the machine selection, the system layout as well as the capacity planning, (ii) AI for the operation of manufacturing systems that require planning and control of the material and the information flows and (iii) digital platforms and ICT technologies for the development and deployment AI applications in manufacturing systems. For each category, the scope and theoretical background are provided and then, selected cases of AI applications are discussed.

Keywords Production scheduling · Manufacturing systems design · Shop floor control · Information flows · Digital Platforms in manufacturing · Supply chain management

4.1 Introduction in the Manufacturing Systems Hierarchical Level

This chapter will focus on AI at the manufacturing system level. A manufacturing system may mean several things, depending on the viewpoint taken. In this work, it is assumed to comprise a combination of machines, cells, intra-logistics devices, and other peripheral devices, used on the factory floor. Moreover, the value chain of connected manufacturing systems, including logistics, is considered within this level. As introduced in Chap. 1, the hierarchical levels of manufacturing systems, equipment, and processes differ from each other in terms of context, and time horizons for making decisions (Fig. 1.6). In more detail, the volume and the variety of

© The Author(s), under exclusive license to Springer Nature Switzerland AG 2023
G. Chryssolouris et al., *A Perspective on Artificial Intelligence in Manufacturing,*
Studies in Systems, Decision and Control 436,
https://doi.org/10.1007/978-3-031-21828-6_4

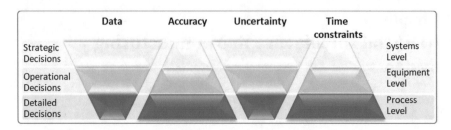

Fig. 4.1 Data, accuracy, uncertainty, and time vs manufacturing hierarchy models

relevant data increase, as data from lower levels are integrated into the system's data structures. Similarly, activities are characterized by a higher degree of uncertainty and stochasticity, with several interdependencies among parameters in non-linear relations. All these inherited attributes make transparency, predictability, and adaptability more challenging tasks for AI. However, at the same time, as uncertainty and stochasticity increase, the need for accuracy of the models' results decreases, at least when compared to the AI models at the lower hierarchical levels and there are longer time frames available for decision making, as shown in Fig. 4.1. With regards to the character of the decisions, they are more strategic in comparison to the other hierarchical levels and can tolerate longer time frames. The two broad application areas to be addressed in this section, comprise the design of the manufacturing system, including the sub-problems of layout design, resources requirement, material handling system, material flow, and buffer capacity, as well as the operation of the system, along with the long-term planning and the short—term dispatching.

4.2 AI for the Design of Manufacturing Systems

"design and planning of a manufacturing system … relate the design to process and the machine selection, system layout, and capacity planning". "In a broader context, the manufacturing system design is particularly concerned with establishing material and information flows, which dictate its architecture." [1].

Manufacturing systems require the combined and coordinated efforts of people, machinery, and equipment, which in turn, requires the appropriate hardware and software infrastructure. Defining a manufacturing system, under the prism of control theory, requires the identification of its inputs, potential states, and outputs (Fig. 4.2). The inputs to a manufacturing system include production orders, raw materials, and labor and can generally be divided into disturbance inputs and controlled inputs. Controlled inputs comprise scheduling, maintenance, and overtime decisions, which the scheduler can regulate within bounds. Disturbance inputs include machine failures and labor outages. The input can be further classified into information e.g. customer demand for the system's products, and material input, i.e. raw materials and energy. The state of a manufacturing system defines the levels for all completed

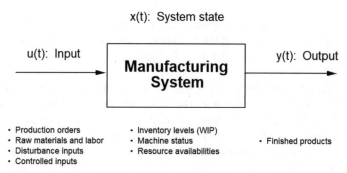

Fig. 4.2 Manufacturing systems from a control theory perspective [1], Reprinted with permission from Springer-Verlag New York INC

and partially completed jobs, the status of all machines (whether active, idle, or under repair), the availability of labor, and the inventories for all materials. The outputs of the manufacturing system may be defined as any portion of the state—for example, the inventory levels of all jobs, ready for shipment on a specified date. The outputs of a manufacturing system can likewise be divided into materials, such as finished goods and scrap, and information, such as measures of the system's performance.

In general, a manufacturing system design can be conceptualized as the mapping of performance requirements, as expressed by values of certain performance measures, onto suitable values of decision variables, which describe the manufacturing system's physical design or the manner of its operation. A performance measure is a variable, whose value quantifies an aspect of a manufacturing system's performance. Performance measures are either benefit measures (the higher the better) or cost measures (the lower the better). Performance measures, similar to the attributes discussed in Chap. 1, can be divided into four categories: time, quality, cost, and flexibility. The definition of the performance measures forms the system objectives. The economic objectives, such as the return on investment, tend to be emphasized most followed by the efficient use of resources and the system's flexibility. What makes economic objectives to be emphasized is that the construction of a manufacturing system is highly capital-intensive. Therefore, no matter how worthy its performance may be in other aspects, a manufacturing system will never be constructed, unless it is shown to be financially viable. The objectives differ from company to company and from case to case. Given the performance requirements, the manufacturing system designer has to create and describe a suitable system design. This design can be captured numerically with the specification of the values of an appropriate collection of decision variables, such as the number of each type of machine, in a manufacturing system, the floor space available, and the existing equipment, which can be incorporated into the new system. Based on the requirements and constraints that are defined, many alternative manufacturing systems are developed and then evaluated, over some

predefined scenarios. A manufacturing system design can thus be viewed as a continuous, cyclical activity, involving the definition of the system's objectives, the development of detailed system requirements and constraints, and the implementation of
the design.

Concerning the difficulty of designing manufacturing systems the reader can get
the first idea from Figs. 1.6 and 4.1. The design of the manufacturing system is an
activity of the highest manufacturing system level where the impact of the decision is more important as it is a matter of strategic planning, whereas there is high
uncertainty. In particular, it is highly due to the inherent characteristics of manufacturing systems. Manufacturing systems are typically dynamic, large, and have a lot
of interacting components. The manufacturing systems are open, influencing their
environment and vice versa. In addition, the analytic formulation of the relationships
between performance measures and decision variables is usually not possible. Data
may be difficult to be measured in harsh processing environments. There are ordi-
narily multiple performance requirements for a manufacturing system, which can
conflict with each other. On top of that, the objectives of a manufacturing system
are often either not well defined at the time that a manufacturing system should be
created, or are subject to a change, which makes design flexibility very important.
Data, regarding manufacturing resources, such as machines and material handling
systems are inexact, especially if the manufacturing process is new. This vagueness
of the inputs to the manufacturing system design process makes its quantitative optimization difficult. This vagueness tends to render futile efforts to hone solutions to
some mathematical optimum.

In the academic literature, the overall manufacturing system design problem
is usually decomposed into sub-problems of manageable complexity and are then
treated separately. These sub-problems are simplified and abstracted with the aid of
assumptions. However, even the simplified problems are usually non-polynomial-
hard (NP-hard), meaning that the time required to find the optimal solution
increases exponentially as the problem size increases linearly. Typical system design
sub-problems are explained hereunder:

- *Resource requirements*—for this problem, the task is to determine the appropriate
  quantity of each type of production resource (for example, machines or pallets)
  in a manufacturing system.
- *Resource layout* is the problem of locating a set of resources in constrained floor
  space. During the resource layout problem, it is considered that the manufacturing
  systems are divided into two areas: 1. the processing area, in which materials are
  processed and individual parts or components are made, and 2. the assembly area,
  in which, if necessary, individual parts or components are joined together in a
  subassembly or final product. The facility layout affects the number of resources
  required because it determines the type of resources being accessible from each
  point in the manufacturing system. The lack of accessibility increases the number
  of resources required. The problem has been formulated in several ways, with
  different degrees of sophistication: 1. the template shuffling formulation, 2. the

quadratic assignment problem (QAP) formulation, and 3. the relationship chart formulation.

- *Material flow* aims to determine the configuration of a material handling system so as for some combination of flexibility, cost, production rate, and reliability of the manufacturing system to be maximized. The material flow decision variables, which must be specified in the design of a material handling system can be divided into two broad categories: those which specify the type of the material handling system and those specifying the configuration of a given type of material handling system. For instance, the travel aisle layout, the number and the locations of the pickup and delivery stations, the pattern of material flow within the travel aisles (unidirectional, bidirectional, or combinations), the number of vehicles required, the routes used by vehicles during specific operations, the dispatching logic used during operation, and the storage capacities of pickup and delivery stations.
- *Buffer capacity problem* is concerned with the allocation of work in process or storage capacity in a manufacturing system. A buffer is a storage space in a manufacturing system for pieces during the processing stages. Buffers serve to decouple the separate processing stages in a manufacturing system since by providing buffer space for inventory among the machines, starvation and blockage are reduced. This comes at the expense of increased inventory.

In industrial practice, trial and error remain the most frequently used design approach. At first, engineers guess a suitable manufacturing system design (guess values for an appropriate collection of decision variables). Subsequently, they evaluate the performance measures of the system. If they satisfy the performance requirements, then, the design process should be stopped. Otherwise, the process will be repeated. The success of the trial-and-error approach heavily relies on the designer or "guesser's" skill, with the help of intuition and the rules of thumb that derive from experience. Researchers have tried to make this process more systematic and enable less experienced users to perform well in the design of structure and operation of manufacturing systems. The methods and tools that have been proposed fall into three broad categories: *operations research, artificial intelligence,* and *simulation.*

Operations research

Operations research requires theoretical expertise and refers to descriptive analytical models and the use of techniques involving Mathematical Programming, which is a family of techniques for optimizing a given algebraic objective function of several decision variables. The objective functions might involve determined (known exactly) or random variables (their probability distribution is presumed). As for the decision variables, they may be either independent of one another, or they may be related through constraints. In the case of, the resource requirements problem, a mathematical programming formulation can explicitly model as constraints in the mathematical program the various limits for design and operating quantities, shared among the resources. Furthermore, mathematical programming has been applied to determine the pattern of material flow in an AGV-based material handling system.

Computer simulation

Computer simulation is the generic name of a computer software class, which simulates the operation of a manufacturing system to provide a set of statistical performance measures (e.g. the average number of parts in the system over time) for its evaluation. The inputs of a computer simulator are decision variables that specify the design (e.g., machine processing and failure rates, machine layout), the workload (e.g., arrivals of raw materials over time, part routings), and the operational policy (e.g., "first come, first served") of a manufacturing system. The simulator assembles these data into a model of the manufacturing system, which includes the rules on how the components of the system interact with each other. The initial state of the manufacturing system (e.g., the number and types of parts initially in inventory at various points in the system) is user-defined and subsequently, the simulator follows the operation of the model over time, tracking events such as the parts' movement, machine breakdowns, machine setups. etc. The optimum decision variables are usually deduced after running multiple simulations. If this search for optimum decision variables is not properly organized, it can become extremely tedious, thus, statistically designed experiments (the process of formulating a plan to gather the desired information at minimal cost, enabling the modeler to draw valid inferences) have been used. As an example, the role of simulation in the resource requirements problem is to determine the bottleneck in the production rate emanating from the "slowest" resource and identify strategies to alleviate bottlenecks e.g. adding additional resources at the bottleneck. Simulation in combination with search-based algorithms has been used in parametric design approaches that seek to supplement the descriptive capabilities of simulation with prescriptive techniques (search algorithms), capable of generating new manufacturing system designs.

Artificial Intelligence-based tools for Manufacturing System Design

Three are the main types of artificial intelligence-based tools that make computers more useful, namely those are search, rules, and neural networks. Search-based tools address the manufacturing system design as a set of values for n decision variables, whereas rule-based tools (also often referred to as expert systems) are built around rules, which consist of an if part and a then part. Neural Networks capture the general relationships among variables, which are difficult or impossible to relate analytically, in a data-driven "black box". Whether the solution approach employs simulation in conjunction with search, neural networks, or other methods, it seems likely that considerable computational resources will be required for realistic industrial problems.

Search methods

In search-based methods, any feasible design can be viewed as a point in an n-dimensional design space. The body of heuristics or intuitively "reasonable" rules, which the designer can use to establish a path through the design space, is called search. A sensible design process should begin at an initial design point, from which a designer seeks to explore the design space, moving from point to point (design to design), and evaluating each point as it arises. Additionally, some sophisticated designers use the information acquired from previous evaluations. At an abstract level, the search methods find solutions by exploring paths. What distinguishes the search methods from each other are the heuristics that are responsible for the exploration to be made. There are several ways of searching for optimal paths, each one with specific advantages. For instance, the branch and bound search is good when the tree is big and the bad paths turn distinctly bad quickly. The branch and bound search with a guess is good when there is a good lower-bound estimate of the distance remaining to the goal. On the other hand, dynamic programming is good when many paths reach common nodes. The A* procedure is considered good when both the branch and the bound search with a guess and dynamic programming are also good.

Rule-based systems

A rule-based system consists of two major components: a rule base and an inference engine. The rule base is a collection of rules that captures human expertise or reasoning in a particular problem domain. The inference engine is a piece of software, which invokes the rules in the rule-base to solve problems. In most rule bases, the rules are interrelated in that the implementation of actions in the "then" part of one rule, may cause a condition in the "if" part of another rule to become true. There are two ways in which the inference engine of a rule base can operate in solving problems. These are referred to as "forward chaining" and "backward chaining". In the "forward chaining", the inference engine answers the question, "Which actions should be taken?". It works by triggering a rule when all the conditions in it are satisfied by the current situation. In the "backward chaining" mode of operation, the inference engine answers the question, "Should a given action be taken?". It works, starting from a rule, whose then part includes the action in question. The application of rule-based systems to the resource layout problem has been limited. Due to the daunting combinatorial nature of the problem, it is impossible to establish generally applicable rules for its solution. However, simplifications should be made. One approach is to divide the resource layouts into a few generic classes, e.g., linear single-row, circular single-row, linear double-row, and multi-row, which are to be mated with one of the two classes of material handling systems: automated guided vehicles (AGVs) or robots.

Neural Networks

The use of neural networks seems promising for the design of manufacturing systems, as their guidance lies in a simple and automatic training process rather than in coding complicated sets of heuristic rules. The design of manufacturing systems may be considered as a generalization application of neural networks and requires a training phase and a use phase. In the training phase, simulations are run to provide sample correlations between the decision variable and performance measure values. A neural network is then trained by exposure to these correlations, which are expressed in the form of training pairs that allows the neural network to generalize the relationship between the decision variables and the corresponding manufacturing system performance measures. Neural network training data for the approach are generated from a limited number of simulation runs in which the decision variables (resource quantities) are varied in some fashion, resulting in a limited number x of {decision variable}–{performance measure} mappings. In most cases, the ratio x/n, where n is the total number of feasible mappings or equivalently the total number of feasible decision variable permutations can be very small and in general, a fashion to reduce n is desired since this would mean an increase of the likelihood of success of the proposed approach.

Resource layout problem

In general, the resource layout problem is an NP-hard class of combinatorial optimization problems [2] and refers to the definition of the workstations' configuration in order for some pre-defined goals e.g. line efficiency [3] to be achieved. In more detail, a large number of feasible design alternatives can be generated, making it extremely difficult for the design engineer to identify a solution that could best satisfy all criteria, which has stimulated the development and utilization of tools for decision support.

The representation of the process components, hierarchy, and sequence, the assignment of tasks to workstations, etc. are some of the problems that are raised when analyzing the resource layout problem [4]. The tools that have been used over the years include simulation, which has supported the generation of the layout, design alternatives [5–7] whereas, neural networks[8, 9], genetic algorithms [10–13], heuristics [14, 15], fuzzy logic [16, 17], and Pareto optimization [18] are some of the most commonly used approaches for the selection of the design alternative. Furthermore, the evaluation of the produced solutions is sometimes challenging, especially when it comes to the achieved flexibility. In this context, Alexopoulos et al. [19] presented a method for the assessment of a manufacturing system's flexibility, based on dynamic programming and statistical analysis of the discounted cash flow estimates of the manufacturing system's lifecycle cost.

Intelligent search algorithms, such as the ones proposed in the [20] are necessary, to enhance the efficiency in exploring the design problem's solution space. Figure 4.3, shows the different steps of a method that has been developed and applied to an automotive case study, as well as the respective functionalities in the form of a workflow (left to right). The methodology comprises two main stages. Stage 1, where the initial configuration is designed. In this stage, an analytical way of calculating the required number of stations and resources is provided on the basis of the product structure and assembly specifications. Stage 2, where the initial design of stage 1 is further detailed, via an intelligent algorithm, capable of selecting specific resources for each station. The selection is based on the individual resource characteristics and the resulting system's performance, estimated with discrete event simulation.

In [21] a method of deriving assembly line design alternatives and evaluating them against multiple user-defined criteria has been presented. Several process requirements, including the process plan and alternative possible solutions, e.g. the joining technology to be used, are provided as input to the AI system (Fig. 4.4). Given this input, the system generates multiple designs and searches the technically feasible one among them. Subsequently, alternative designs are evaluated with multiple criteria,

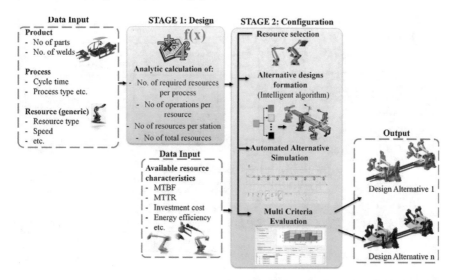

Fig. 4.3 Multi-criteria assembly line design and configuration [20], Reprinted with permission from Elsevier

such as investment cost, availability, equipment reuse, production volume, and flexibility. As a result, suitable alternative cell designs are generated and proposed to the designer (Fig. 4.5). The method discussed can translate a cell design decision making problem into a search one, and it can, therefore, be addressed more systematically, whereas the enhancement of the solution is made by using AI. The method was implemented in the form of a decision support tool, capable of identifying good quality solutions.

The performance of the designed line, in terms of possible bottlenecks, should be investigated before setting up the designed line. This is usually performed using discrete event simulation. As an alternative, a data-driven algorithm based on the ARIMA method has also been suggested [22]. Other attributes that can be evaluated are the inter-cellular movements and the utilization of machines. Noktehdan, Karimi, and Kashan (2010) applied a grouping version of a differential evolution algorithm and its hybridized version with a local search algorithm for the evaluation of cell formation alternatives over the two aforementioned attributes [23].

Besides the aforementioned evaluation criteria, ergonomics has also been accounted for in many approaches. Arkouli, Michalos, and Makris (2022) reviewed a wide range of tools that are used for the assessment of physical and mental workload in manufacturing applications and proposed a multi-criteria decision-making methodology to select a data collection technique based on the particularities of each manufacturing case [24]. In industry, it is quite common that there exists ergonomic knowledge that may not be applied in the early product development process, given that it is not integrated into virtual/design tools. This causes problems since ergonomics risks are identified at later development stages, resulting in higher costs [25]. As a countermeasure, researchers have proposed decision support tools

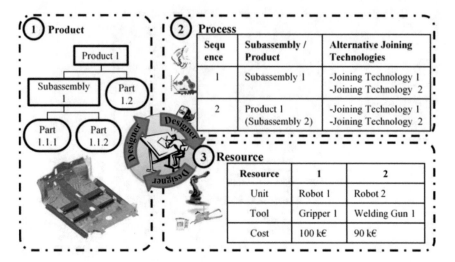

Fig. 4.4 Alternative designs for product design [21]

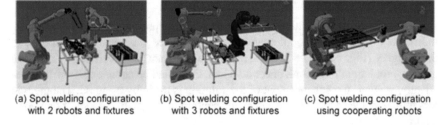

(a) Spot welding configuration   (b) Spot welding configuration   (c) Spot welding configuration
    with 2 robots and fixtures   with 3 robots and fixtures   using cooperating robots

Fig. 4.5 Alternative cell designs created through a generative design process [21]

that help in identifying risks or assessing the design of a manufacturing system. For instance, Alexopoulos et al. (2013) presented the development of ergonomic evaluation methods integrated with VMS and DHM named ErgoToolkit, which enabled early ergonomic analysis with the use of dynamic human task simulation [26]. The approach along with the common design tools uses a rules-based system to define sets of postures and employs an inferencing logic to decide on the type of posture selected for the execution of a task. The tool may be used by non-ergonomic experts (e.g. production or simulation engineers) to get rough ergonomic assessment and proceed to the necessary production process changes early in the design phase. Moreover, it is relatively easy for the ergonomic rules, included in ErgoTookit, to be tailored to company-specific needs. For example, the set of invalid postures.

Material Handling Systems—AI for Automated Guided Vehicles and Autonomous Mobile Robots

Assembly systems are usually organized in lines that involve several stations, where various parts and components e.g. screws, clips, cables, etc. are used in order to make the final products. These parts are stored in boxes, pallets, etc. located at specific positions in each station. The replacement of any of the assembly station boxes that are under depletion is performed using boxes that exist in large warehouses (markets) outside the line. The mass customization trend has stressed the mixed model assembly paradigm, where different quantities of parts are consumed by the different models. This in turn, frequently results in unbalanced inventory levels within each station and the need for consumables and parts to be supplied to the stations dynamically. Whenever there is a need for one or more boxes to be replaced, multiple alternative solutions may be realized. The material-handling system has to move parts from one machine to another in a timely manner. Additionally, enabling the exploitation of the Flexible Manufacturing Systems (FMS) benefits requires the seamless movement of parts from station to station, which can be achieved with several different types of material-handling systems.

The selection of the type of a material-handling system is a function of several system features, such as the load and bulk of the part and perhaps the part fixture. For instance, large, heavy parts require large, powerful handling systems, such as roller conveyors, guided vehicles, or track-driven vehicle systems. The number of machines to be included in the system and the layout of the machines also present another design consideration. If a single material handler is to move parts to all the machines in the system, then the work envelope of the handler must be at least as large as the physical system. A robot usually can reach one or two machines and a load-and-unload station, whereas a conveyor or automatic guided vehicle (AGV) system can be expanded to include miles of the factory floor.

AGVs and autonomous robot technology are important solutions in the field of handling systems that can be used for part supply in assembly lines, given their ability for efficient handling and transport. Nevertheless, their integration into production systems comes with challenges in the planning and coordination of their functioning. There are different alternative pathways that a mobile unit may travel on when considering possible part supply operations. These part supply operations can be clustered into tasks for each agent. For instance, moving to target locations, identifying pallets, boxes, etc., but also loading and unloading. Moreover, precedence constraints among the different operations should be considered, e.g. an AGV or mobile robot cannot go to the station for the unloading of a box before this box is loaded from the respective market.

Kousi, Michalos, et al. (2016) proposed a decision-making method for the plan generation of part supply operations for Mobile Assistant Units (MAUs) that used time and inventory levels as decision-making criteria. An efficient part supply plan should account for a number of constraints, including the storage areas or shelves, where the boxes are located during their transportation. Since different types (dimensions) of boxes are used to store the different parts, each storage area/shelf of the mobile unit can store only specific combinations, in terms of the number of boxes. Each feasible combination of boxes that a mobile unit can carry simultaneously in all its storage areas/shelves is defined as a configuration. Given the predefined dimensions of each type of box and the characteristics of each mobile unit, the total number of feasible configuration alternatives for each mobile unit can be calculated. On the shop floor, the existing mobile units may have varying specifications with respect to their capacity, the number of shelves, speed, dimensions, etc., meaning that the total number of configuration alternatives will be different for each mobile unit. A crucial constraint that should be taken into consideration is that at least one of the mobile unit's shelves needs to carry at least one box.

Based on this formulation, a two-step approach has been proposed; the first step is the calculation of the number of task alternatives for each box type allowed in each configuration. The second step is the calculation of the overall number of task alternatives, which is made by aggregating the product of the number of the alternative tasks for each type of box, allowed in each configuration. Considering that the mobile unit may have one of the abovementioned configurations, alternative configuration combinations can be formed as a tree (Fig. 4.6). The tree is constructed in steps, starting with the identification of the number and type of boxes to be considered in the planning.

In layer 1, a mobile unit is selected among any of the available mobile units. For each layer, either sequential or randomized selections are made, having no impact on the formulation of the alternative. All possible configurations are listed as a tree's branches. For each of the remainder mobile units, all possible configurations are listed in separate layers by considering the configuration/boxes of the previous layer

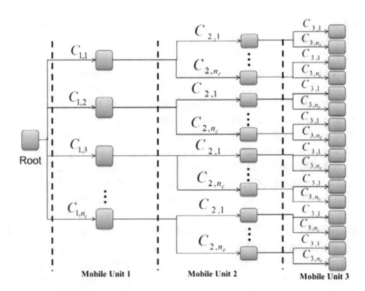

Fig. 4.6 Short-term planning for part supply—Configuration alternatives tree [27], Reprinted with permission from Elsevier BV

to exclude any duplication of tasks in the same tree. In the configuration alternative tree, each node can be replaced by the tasks that are compatible with each configuration, meaning that the resulting tree can be further expanded, according to the number of tasks that have to be carried out. The selection of an alternative among the set of feasible ones requires the definition of metrics for the quantification of the performance of each one of them. In this view, transportation—representation of the time required by the mobile unit to perform the assigned tasks has been selected, as a cost criterion. Its minimization will result in the maximization of the utilization of resources as well as in the reduction of idle time.

A clonal selection algorithm has been proposed for the minimization of the distance travel of automated guided vehicles, in material handling operations, and in enhancing the performance of an FMS [28]. Moreover, genetic algorithms have been proposed for the redundant configuration of robotic assembly lines to address stochastic failures [29]. On the other hand, Kulak (2005) developed a decision support system that relied on axiomatic design principles, the fuzzy information approach for incomplete information, and a multi-attribute decision-making approach to choose between equipment types such as conveyors, industrial trucks, AGVs, cranes, etc. [30].

4.3 AI for the Operation of Manufacturing Systems

The operation of a manufacturing system requires planning the material and information flows in it. The production of products on time and in sufficient quantities is a direct consequence of the system's information flows; commanding information from human planners or planning software prescribes the material flow in the system, while the sensory information monitoring the status of the system's resources is used to deciding on the appropriate commands. Therefore, determining the commands, which prescribe the material flow in the system, is a fundamental activity in the operation of a manufacturing system.

In industry, the operation of a manufacturing organization is typically divided into three hierarchical levels: strategic planning, operations planning, and detailed planning and execution [1] (Fig. 4.1). At the strategic level, few decisions are made by the top management of the organization, but each decision takes a long time, whilst its impact is felt throughout the organization. For example, which manufacturing system structure (flow line, cellular system, job shop, project shop, or continuous system) is the most consistent with the organization's overall strategy, the entry into a new market, or the acquisition of additional manufacturing capacity. Finally, performance measures, such as return on investment, market share, earnings, growth levels, and so forth, are established for the organization.

Concerning the distinction of the types of commands, there are two main types: high-level and low-level. High-level commands—called long-term planning and hereafter referred to as planning—are concerned with determining the aggregate timing of production. Planning dictates the flow of materials into the manufacturing system and determines its workload. Low-level commands—called short-term dispatching and hereafter referred to as dispatching—are concerned with the detailed assignment of operations to production resources. Dispatching, dictates the flow of materials, within and out of a manufacturing system, and determines which production resources are to be assigned to each part's operation and the time that each operation is to take place. Since low-level commands control individual operations, they should be generated much more frequently than high-level commands, which is also aligned with the illustration of data availability and time constraints in the different manufacturing levels of Figs. 1.6 and 4.1. Indicatively, the time between commands is in the order of seconds or minutes. Long-term planning and short-term dispatching are collectively labeled, in broad terms, as "production scheduling." The scheduling problem in most production environments is stochastic and dynamic. Therefore, the scheduling methods either give some guarantee as to the insensitivity of the schedule to future disruptions or explicitly reflect the uncertain nature of the available information.

Common approaches dealing with the production scheduling problem include operations research techniques (more specifically, mathematical programming), as well as AI techniques. AI techniques include rule-based systems that seek to capture generic scheduling rules, applying to a wide range of situations. Knowledge in a rule-based system can be classified into static (or data) and dynamic knowledge (or solution methods). Static knowledge includes all information about the manufacturing system itself (e.g., the number and types of machines) and the production objectives (e.g., the part types to be produced, along with their processing sequences, quantities, due dates, etc.). The dynamic knowledge describes available expertise to the way that feasible schedules can be derived, and consists of theoretical expertise, which refers to operational research techniques that deal with the management of time and resources; empirical expertise, which consists of heuristic dispatch rules; practical dedicated expertise, which is provided by experienced shop floor supervisors, who are aware of the constraints that have to be taken into account, when preparing the schedule. In a rule-based approach, these types of expertise are captured in the knowledge base in the form of rules.

The challenges that are encountered in scheduling are related to the high variability of the actual manufacturing systems. The entities to be scheduled are typically referred to as jobs, where each job corresponds to an individual part. A job comprises individual production operations, which are referred to as tasks. Classical scheduling problems can be categorized based on the following dimensions:

- *Requirements generation*: requirements can be generated directly by the customers' orders, or indirectly by inventory replenishment decisions. This distinction is often made in terms of an open shop (jobs are upon customer request and no inventory is stocked) versus a closed shop (the customer requests are serviced from the inventory and the jobs are generally a result of inventory replenishment decisions).

- *Processing complexity*: Processing complexity is concerned with the number of tasks, associated with each job.

The scheduling criteria include common measures for schedule performance, namely the utilization level of the production resources, the percentage of late jobs, the average or maximum tardiness (i.e., the positive part of the difference between a job's actual completion time and its desired completion time) for a set of jobs and the average or maximum flow time (i.e., the difference between the completion time of the job and the time at which the job was released to the manufacturing system) for a set of jobs. On the other hand, the scheduling environment deals with assumptions about the certainty of information, regarding future jobs and can be classified as being static or dynamic. In a static environment, the problem contains a finite set of fully specified jobs.

In operational planning, the objectives of the strategic manufacturing plan are converted into more detailed and specific plans. One of the most important of these plans is the Master Production Schedule (MPS) expressed in specific product config-urations, quantities, and dates, which is frequently the key link between top manage-ment and the manufacturing facility. It is also a key link between manufacturing and marketing and serves to balance the demands of the marketplace against the constraints of the manufacturing facility. The master production schedule is one of the inputs to material requirements planning (MRP), which obtains future require-ments for finished products to generate the requirements for all the sub-assemblies, components, and raw materials that go to make up the finished product. Although MRP is a generic methodology, due to the vast amount of data processing when many part types with possibly many layers of subassemblies are involved (Fig. 4.1), it is usually handled by computer software, called the MRP system.

At the execution level, many decisions are made, each requiring a much shorter time, e.g. capacity planning. Although the impact of each decision is local in time and place, the great number of decisions, taken together, can have a significant impact on the organization's performance (Fig. 1.6). Focusing on capacity planning, it should be mentioned that it is responsible for providing projections of the capacity needs, implied by the material plan so that timely actions to be taken to balance the capacity needs with available capacity. Flexibility is one of the attributes that should be considered in decision-making. Therefore, Alexopoulos et al. (2010) suggested a method for the comparison of the performance flexibility of manufacturing systems, in uncertain environments, by considering the capacity planning constraints [31].

The shop floor control system is responsible for detailed planning, such as the allocation of resources in the forms of labor, tooling, and equipment to the various orders on the shop floor or determining the priority sequence of orders in a queue, and monitoring the execution of the detailed plans, such as capacity utilization at the machining center (i.e. input/output control). The shop floor control system has to allocate four major resources available to it: (a) material, (b) labor, (c) tooling and (d) machine capability. Three major activities make up a detailed assignment. These are (1) order sequencing/dispatching, (2) scheduled maintenance, and (3) other assignments. In actual manufacturing systems, dispatching is typically performed ad hoc, or via the application dispatch rules. Decision-making procedures may provide a comprehensive, fundamentally sound alternative to empirically stated dispatch rules. In making resource-task assignment decisions, they can consider resources and tasks simultaneously and can assign each task to a specific resource, in contrast to the dispatch rules, which only select the next task to be performed.

The shop floor control system collects data from the shop floor and directs a flow of information back to the operations planning level. It uses information provided by the planning systems for the identification of those orders and the action to be taken and complements other planning systems, such as material requirements planning and capacity requirements planning. These planning systems, provide the resources required by the shop floor control system and set the objectives to be achieved by this system. The Shop floor control is then responsible for using these resources for its objectives to be achieved in an effective and efficient fashion. The shop floor control is supposed to close the loop between the planning and execution phases of the manufacturing system by feeding information from the shop floor back to the planning systems.

After interpretation and aggregation, this information will pass on from the operational level to the strategic. Figure 4.7, depicts the major components of the manufacturing planning system. The figure depicts two flows. The first flow is that of the product and the attendant physical allocation of resources. The second flow is that of information. As the shop order progresses through the various stages of processing, it generates information, which is then used in monitoring its progress resources being:

Fig. 4.7 Major Components of the Manufacturing Planning System [1], Reprinted with permission from Springer-Verlag New York INC

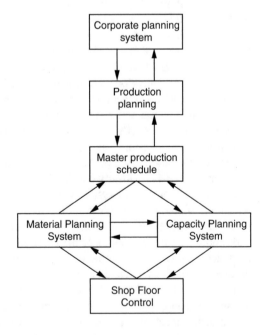

- *Manpower*. This resource includes all of the personnel that the shop floor can draw on for the execution of the plans released to it (overtime, workers transferred in from other locations, part-time help, and multiple shift operations).
- *Tooling*. It refers to all of the equipment and special fixtures that are used during the setup and the operation of a machine or the assembly operation.
- *Machine Capacity*. This is the total amount of productive capacity, offered by the equipment available.
- *Material*. This is the total stock of components that can be used in completing shop orders.

Information links the planning system with the execution system. The information, provided by the shop floor control system is the major means by which the planning system can track the physical flow: e.g. the current location of the shop order, the current state of completion, actual resources used at the current operation, actual resources used at preceding operations, any unplanned delays encountered. Corrective action by the management is required at any time that the actual progress of a shop order exceeds some predefined margin of difference from its planned progress. The progress can be monitored along several dimensions: stage of completion, costs, scrap produced, or nearness to the due date to name a few. Capacity control refers to any corrective actions that attempt to correct the problems by means of very short-term adjustments at the level of resources, available on the shop floor e.g. changes in work rate, use of overtime or part-time labor, use of safety capacity, alternate routings, lot splitting, subcontracting of excess work.

The final step in production is order disposition (which can be either order completion or scrap). The quantity received from the shop floor is recorded and the performance of the shop floor system is evaluated, based on measures such as the number of labor hours required, the breakdown of labor hours between regular time and overtime, the materials required by the order, the number of hours of setup time required, the amount of tooling required, the order's completion date, the amount of rework or scrap generated by the order, the number of machine hours required.

In product-oriented Industrial Product Service Systems (IPSSs), the customers benefit from the combination of a product, which is accompanied by a set of functionalities and services. The IPSS supports the provision of services that can be offered by the product manufacturer. The services can offer a wide range of functionalities, ranging from ensuring the product's original functionality to augmenting the product's original functionality. A company's shifting to IPSS, poses many challenges, such as changing its business model. One of the most important challenges for the establishment of IPSS is the appropriate planning of the resources for production, deployment, and installation into the customers' site. In Alexopoulos et al. (2017), a multi-criteria resource planning method and tool for optimizing the production, delivery, and installation of IPSS, has been developed [32]. The solution employs the AI technique for the generation of the alternative IPSS's production and installation plans and evaluates them on performance measures for production and installation, namely time and cost. Moreover, through the integration of the planning tool with

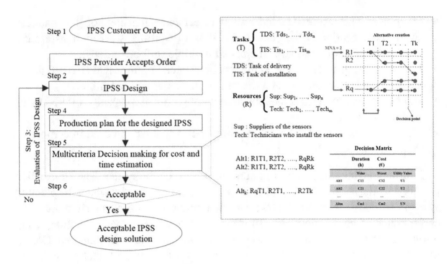

Fig. 4.8 IPSS production and installation planning flowchart (left); intelligence search for evaluating different options (right) [32], Reprinted with permission from Springer

the IPSS design phase, information for generating the Bill of Process and Materials is presented. The objective of the AI planning method is to find an optimal solution that decides what IPPS equipment (e.g. sensors) suppliers to select, which resources (e.g. IPSS service installation technicians), and when they should perform what processes/tasks at the IPSS provider or customer site. The planning method proposed in this study is based on the approach proposed in [33] and it defines the approach of assigning a set of resources to a set of tasks, under multiple and often conflicting optimization criteria. Figure 4.8 illustrates the workflow of the proposed methodology as well as the main principles of the intelligence search approach that is used.

Service-oriented architecture and a heuristics-based approach have been suggested for the dynamic scheduling of robots for the material supply [34] and implemented as shown in Fig. 4.9.

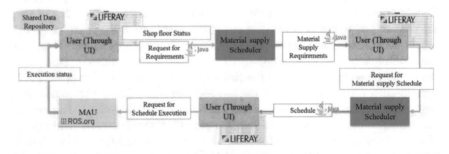

Fig. 4.9 SoA implementation [34], Reprinted with permission from Elsevier BV

A decentralized multi-agent system framework and a distributed version of the Hungarian Method have been proposed for mobile robot scheduling [35]. An adaptive preferences algorithm has been utilized for the identification of the optimal scheduling policy, in the multi-robot assembly of large structures (aerospace) [36]. Chen et al. (2011) investigated generalized stochastic Petri Net models in tandem with the Monte Carlo method and multi-objective optimization for scheduling the human–robot collaborative assembly of an industrial power supply module [37]. Re-scheduling has been performed for a team of mobile robots by using a combination of the Critical Path Method, a resource leveling method, and a monitoring agent for the initialization of the re-scheduling process [38].

Dispatching

Dispatching is concerned with the detailed assignment of operations to production resources. The decision-making for shop-floor dispatching and control is responsible for preventing the deterioration of the manufacturing plan's execution. Traditionally, much of the research done in dispatching has focused on heuristics, called dispatch rules. A dispatch rule is a method of ranking a set of tasks, which are waiting to be processed on a machine. The task with the best rank is selected to be processed on the machine. A schedule results from the rule's repeated application, whenever a machine is available for assignment. When more than one machine is available at a given time, a machine selection rule is also required for the order's specification for which the machines will receive assignments.

Because dispatch rules are heuristics, no dispatch rule has been found to perform best in all situations. Factors that affect the suitability of a dispatch rule include the distribution of the processing times of the tasks to be dispatched, the distribution of the job due dates, the distribution of job arrivals over time, the performance measure to be optimized (e.g., mean tardiness, production rate, etc.), the nature of the jobs' process plans (presence of assembly tasks, number of tasks per job, etc.).

The dispatch rules have a lot of limitations; for example, they only prescribe which task to be assigned and not which resource the task should be assigned to. Therefore, when several resources become available at the same time, they should be arbitrarily assigned to the tasks, selected by the dispatch rule or another rule for machine selection has to be applied. The dispatching rules, partition the dispatching problem accordingly by first considering the resources and then the tasks. The corresponding performance measures can be based on completion times, (due dates, mean lateness, etc.) or based on inventory and utilization of costs e.g. mean number of jobs waiting for machines. Heuristics have been used in a set of approaches [39–42]. Other approaches for dispatching include Dynamic Programming Approach, as well as probabilistic approaches [33], iterative backwards and forwards scheduling [43], extreme value theory [44], and neural networks[45]. The progress plot of standard PERT network data is an approach to distinct between tolerable deviations from the plan and developments that may call for remedial action.

Older approaches on scheduling, some of which are still relevant, involve genetic algorithms [46]. Most job shop scheduling methods, reported in the literature, usually address the static scheduling problem. These methods neither consider multiple criteria, nor do they accommodate alternate resources to process a job operation. The genetic algorithms approach is a schedule permutation approach that systematically permutes an initial pool of randomly generated schedules so as to return the best schedule found to date (Fig. 4.10). A dynamic scheduling problem was designed to closely react to a real job shop scheduling environment. Two performance measures, namely the mean job tardiness and the mean job cost, were used to demonstrating multiple criteria scheduling. To span a varied job shop environment, three factors were identified and varied between two levels each. The results of this extensive simulation study indicate that the genetic algorithms scheduling approach, produces better scheduling performance in comparison with several common dispatching rules.

Decision making for scheduling has been addressed with the coupling of digital (discrete event simulation) and physical (intelligent analysis of huge amounts of information) worlds [47], the artificial immune system algorithm [48], and the branch-and-bound algorithm [49], but also reinforcement learning [50, 51] game theory [52] and the hybrid backward-scheduling method with hierarchical finite capacity shop-floor models and discrete simulation [53]. Decision-making for nesting and scheduling has been treated with heuristics-based search approaches [54].

The MADEMA approach has been proposed for dynamic dispatching, during the operation of the manufacturing system [55]. Within each work center, whenever one or more resources become free after the completion of their tasks, a dispatching decision takes place by assigning one pending task to each free resource. Simulating this dispatching function over time results in a list of assignments for each work center.

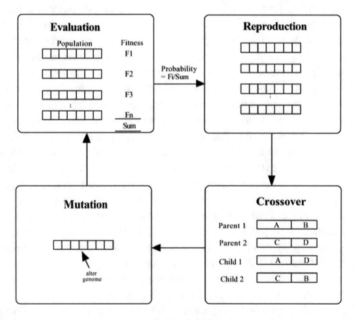

Fig. 4.10 A generic genetic algorithm, [46], Reprinted with permission from Kluwer Academic Publishers (Boston)

The schedule for the entire manufacturing system is built up by combining the assignment lists for the individual work centers. Dispatching has to be performed dynamically because the task arrivals and the resource breakdowns, constantly change the tasks and the resources involved. If dispatching decisions had been performed ahead of time, then unforeseen interruptions, such as breakdowns would have caused many of the resulting assignments to become infeasible. Since this dispatching should result in feasible assignments, it is a finite-capacity process.

The problem of scheduling the dyeing of fibers, the yarn spinning, and the carpet weaving operations of a typical textile plant, from a decision-making perspective, is presented in [56]. In this approach, the product items ordered are grouped and then the duration of the weaving task, for each group, is roughly estimated via heuristic rules, without determining the nesting layout of the proposed schedule. In many industrial cases, both the nesting and the scheduling problems have to be addressed at the same time.

Mourtzis, Doukas, and Vlachou (2016) proposed a knowledge-enriched, short-term job-shop scheduling mechanism, implemented into a mobile application [57] (Fig. 4.11). They focused on the short-term scheduling of the machine shop's resources, through an intelligent algorithm that generated and evaluated alternative assignments of resources to tasks. Based on the requirements of a new order, a similarity mechanism retrieves successfully past executed orders, together with a dataset that includes the processing times, the job and task sequence and the suitable resources. In addition to that, the similarity mechanism is used for the calculation of the orders' due-date assignments, based on the knowledge stored in past cases. Afterwards, it adapts these parameters to the requirements of the new order so as to evaluate the alternative schedules and identify a good substitute in a timely manner. The derived schedule was presented on mobile devices and it was possible to be manipulated by the planner on-the-fly, respecting the tasks' precedence constraints and machine availability.

Giannelos et al. (2007) studied dispatching policies, through the prism of the chaos and nonlinear dynamics concepts [58]. The scheduling of a simple manufacturing system, with the help of common assignment rules, has been simulated. The results were studied and analyzed with the help of time-delay plots. The method was tested against conventional rules. The use of chaos-related concepts, such as phase portraits and time-delay plots, reveal interesting geometric patterns of the variables, associated with the production scheduling problem and may often bring to light orderly structures. A similar analysis of other scheduling variables or interrelationships between variables, e.g. flowtime and tardiness, may also reveal orderly patterns.

Dynamic job rotation has been addressed with mathematical programming and heuristic solutions [59], whereas the layout of optimal job rotation for the upper bank of an assembly line is shown in Fig. 4.12. The job rotation enables production systems to cope with the fluctuating market demand by exploiting the benefits of flexible workforce. It provides employees with a more engaging working environment, resulting in far less monotonous and repetitive tasks. There is a study on a dynamic job rotation tool, which enables the efficient allocation of assembly tasks to suitable operators, at any point of time, leading to more balanced workload distribution and thus, having a 'dynamic line balancing' achieved. A hierarchical approach to multiple criteria

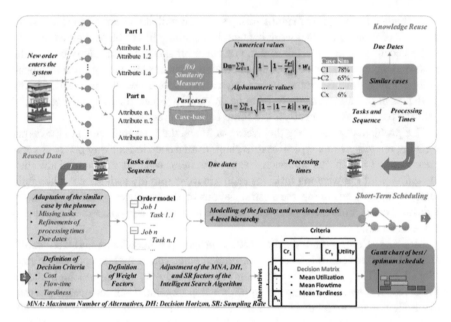

Fig. 4.11 Workflow of the knowledge-enriched short-term scheduling (KES) method [57], Reprinted with permission from Springer-Verlag Berlin/Heidelberg

and decision-making algorithms were used for the implementation of the tool. The tool generates alternative rotation schedules and evaluates them against predefined criteria. A monitoring system coupled with a database, was also used for the storage of the required data for the generation of the rotation schedule. This monitoring allowed the triggering of the rotation schedule generation, in case that the production deviates from the existing plan in unexpected occurrences (delayed assembly, equipment breakdowns, etc.). This enabled the system to absorb any disturbance without disrupting the production process. Fatigue levels have been considered for the rotation triggering.

Neural Networks have been proposed by Hao, Lai, and Tan (2004) to address the personnel scheduling problem in place of integer programming-based models and heuristics, which have been typically used [60]. In view of addressing this problem, modelling and simulation of workers competence, skills and preferences have been studied by a plethora of researchers [61–63]. Zülch, Rottinger, and Vollstedt (2004) have shown three different approaches for personnel planning and re-assignment by exploiting simulation capabilities, in order to consider the plurality of possibilities for personnel assignment and to exploit the flexibility of human resources [62]. The preferences and ability restrictions of the existing personnel have been considered. In their case study, a re-organized personnel structure, has resulted in a significant improvement of the logistical goal achievement. Personnel absenteeism was dealt with the mode of action of a dynamic priority rule, which controlled potential bottlenecks, with respect to specific functional elements during the simulation run.

Fig. 4.12 Layout of optimal job rotation for the upper bank of the assembly line [59], Reprinted with permission from Elsevier

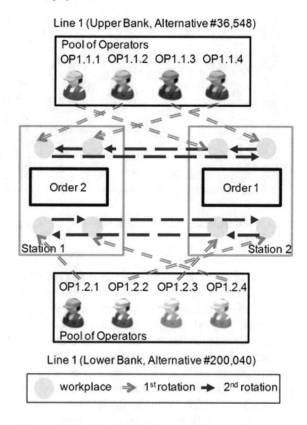

As for the integration of scheduling approaches, a web-based tool that uses an intelligent search algorithm for multiple criteria decision making has been demonstrated [64]. Through a user-friendly web interface, production engineers can represent assembly lines, the tasks to be performed for each product and the operator's characteristics. The web-based tool generated job rotation schedules for human based assembly systems based on an intelligent search algorithm used for the generation of alternative solutions to the scheduling problem. Multiple criteria decision making is used for the evaluation of the job rotation schedule alternatives, according to criteria having derived from industrial assembly line requirements. Figure 4.13 depicts the architecture of the dynamic job rotation tool.

The quality of rotation schedules was quantified by means of the following criteria: (a) competence i.e., the ability to perform an assembly task without making any mistakes, (b) the operator's fatigue accumulation, i.e. the amount of fatigue induced on an operator, (c) fatigue distribution between operators, (d) distance travelled, (e) travelling distance distribution between operators, (f) cost, i.e. operators with high training/expertise and consequently, higher cost rates, should be efficiently assigned, (h) repetitiveness. The method uses the calibration method for the selection of the values of these decision parameters provided in [40].

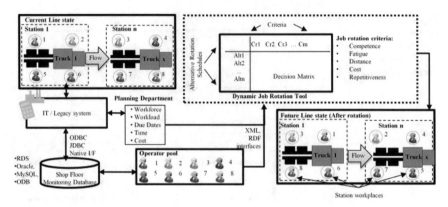

Fig. 4.13 Dynamic job rotation tool architecture [64], Reprinted with permission from Edition Colibri AG

Efthymiou, Pagoropoulos, and Mourtzis (2013) implemented a smart scheduling method into a virtual platform and applied it to a refrigerator factory [65]. The method employs the modelling of the factory's resources and the assignment of the workload of the resources in a hierarchical fashion. The developed software system simulates the operations of the factory and provides a schedule for the manufacturing system's resources. The system is integrated with a holistic virtual platform, namely, the Virtual Factory Framework that enables the exchange of data, related to product, process, resources, and key performance indicators along with other software components that are also integrated with the Virtual Factory Framework.

Kousi et al. (2019) investigated the scheduling of autonomous mobile robots for material supply, using discrete event simulation and a multi criteria approach [66]. The authors have also presented the design and prototype implementation of a service-based control system, responsible for the material supply operations planning and coordination in assembly lines. The material supply processes are carried out by autonomous mobile units that are responsible for the transportation of the consumables from the warehouse to the production stations. The plan generation, based on time and inventory level driven criteria, is automatically carried out by a web-based software that can also distribute the derived plan to the autonomous mobile units. The proposed system has been implemented on a case inspired by an actual production line from the automotive assembly sector. Discrete event simulation has been employed for the investigated production system, to derive the specifications for the mobile units (e.g. number of boxes that can carry simultaneously) that may serve the system efficiently. The results indicate that the proposed architecture, integrated with the discussed mobile assistant units, may provide high quality solutions, with respect to the end user's criteria.

The proposed SoA schema (Fig. 4.14) comprises three levels, namely the (1) Decisional, (2) Execution Control and (3) Physical Execution level. The Decisional and the Physical Execution levels are connected through the Execution Control level, which is responsible for the decentralized integration and communication of the

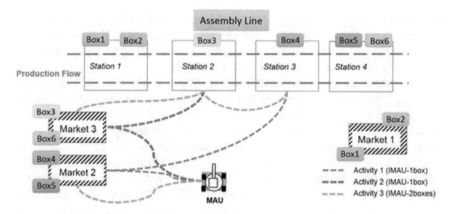

Fig. 4.14 Different pathways of Mobile Autonomous Units (MAU) [66]

system's individual components. The Decisional level delegates the interpretation of task assignments to the Execution Control level, which in turn, delegates the execution of the tasks to the Physical Execution level.

The implemented shared data repository is responsible for the storage of multiple pieces of information, provided by the MES. Such information includes the current activities, dispatched in the shopfloor as well as their execution status. The inventory levels at each point of the execution are also stored into the repository. The information stored into the shared data repository is modelled by an ontology that defines a set of ontology classes, object and data properties. The shop floor ontology corresponds to the entire assembly plant and includes the assembly lines, the warehouses for the storage of the boxes with consumable parts and the available MAUs. Each assembly line consists of several assembly stations, where several boxes are stored. Respectively, each of the warehouses consists of warehouse shelves where boxes with consumables are stored there. Siatras et al. (2022) presented a toolbox of both model-based (mathematical programming) and data-driven (artificial neural networks) agents for scheduling the paint shop environment of a bicycle industry [67].

Control

Approaches to the control of manufacturing systems being available in literature, among others, include the coupling of the physical and digital worlds together with discrete event simulation [47], distributed machine control systems along with a multi-attribute objective adaptive approach and the calculation of the mean-squared deviation of job completions [68], the analytic hierarchy process based approach [69], as well as the discrete event simulation together with the agent-based approach [70], a self-learning method, based on the probability distribution of random events [71]. In general, agent-based methods have been commonly used for control as an alternative to centralized systems [72]. Tang and Wong (2005) have proposed multi-agent based control, including reactive agents for high robustness, high responsiveness and expandability [73], the multi-agent based control with operation based time-extended negotiation protocol for on-line adaptation was investigated in [74]. Mahesh et al. (2007) suggested a web-based multi-agent system for geographically distributed agents, interpreting commands from one another [75].

Disruption monitoring in a manufacturing system, is discussed in [76]. A prototype implementation, using an immune based ontology and multi-agent systems, is suggested in order for decision makers to be provided with recommendations on reaction strategies. A distributed data storage mechanism collects and provides real time data to an ontology model, using IEC61499 function blocks, for reconfigurable resources. Moreover, decision making is enabled via Semantic Web Rule Language (SWRL) rules, allowing the adaptive task execution among the manufacturing resources, in case of a break-down event. RFID data and wireless information have favoured the disruption monitoring as well as the deployment of Just-In-Time manufacturing [77]. Torres (2017) proposed an intelligent system for text recognition in the industry and for objects identification, using an edge-enhanced maximally stable extremal regions (MSER) algorithm [78]. This study has focused on the automatic detection and recognition of text in unstructured images for use on shop floor mechatronic systems with vision systems, for the identification and recognition of patterns in products, thus reducing the need for predefined calibration models. The confirmation of undisturbed production or the detection of disruptions are the input of the shop floor control that in turn, triggers corrective actions.

George Michalos, Sipsas, et al. (2016) investigated a control framework, enabling the real time re-configuration of shop floors with autonomous production units [79]. In particular, they considered (a) autonomous mobile manipulators capable of repositioning themselves on the line, (b) flexible grippers. Each one of these elements allows a different reconfiguration degree. Reconfigurable tools enable the easy adaptation of the production process to disturbances and market variations. The degree of autonomy can be further increased when combining such tools with mobile robotic platforms, since these robots can undertake multiple roles within the assembly system, allowing it to recover smoothly from many technical problems. Cooperating robots, on the other hand, i.e. robots communicating with each other for carrying out common tasks, involve workspace sharing, motion synchronization, program synchronization, and linked motion. Typically, Flexible Manufacturing Systems (FMS) comprises the application of agent-based control in Computer Numerical Controller (CNC) machines. In this case, the machines have several programs stored in and the agents decide which one to be executed, depending on the pending operations. The dynamic nature of the tasks requires more complex coordination among the resources (horizontal integration) as well as a higher level of coordination services (vertical integration). Figure 4.15 illustrates the hierarchical architecture and the decisions to be made for each level.

The required functionalities for the application of the control logic require 1. monitoring and data collection, i.e. collection of signals from sensors, MES systems, PLC, or services running on the controllers of each resource, to identify events that may impose the need for line reconfiguration, 2. event detection and analysis that allows to traced back to the information regarding the malfunctioning resource,

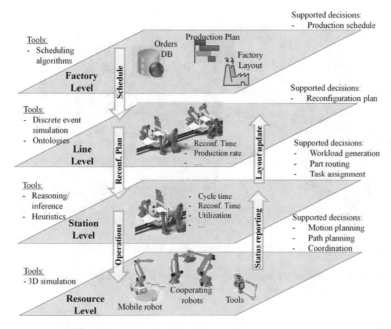

Fig. 4.15 Reconfiguration of the production system at the Resource, Station, Line, Plant levels [79], Reprinted with permission from Pergamon

the station that the resource belongs to, the processed product and the ongoing or disrupted task to the error signal, 3. the selection of the reconfiguration strategy: (a) reassignment of the disrupted task to another station, (b) the use of mobile units as replacements for malfunctioning units or as additional resources within a station.

Figure 4.16 illustrates decision-making based on the proposed approach, whereas each one of the steps are detailed hereunder:

- *Unit Suitability*: check whether the mobile unit is capable (docking station compatibility, robot capabilities, etc.) of performing the task that failed in the station. If there is no compatibility between the operation and the unit, the decision- making stops, otherwise it continuous to the next steps.
- *Unit Availability*: check whether the unit is at rest and can directly start moving to the requested area, whether the unit is in use or not available for a short period, or whether the unit cannot be used as it has to work in another area. The decision-making stops in this case.
- *Tool suitability*: the test concerns the suitability of the tools which the robot is equipped for the requested operation. If the unit has no tool attached to it or it has a non-suitable tool, a decision as to where a suitable tool can be found needs to be taken (see next step), otherwise, the next step for finding a tool can be skipped
- *Nearest Free Suitable Tool*: the control logic examines the location of the mobile unit concerning the last known locations of suitable tools and determines which locations are candidates for tool pick up.

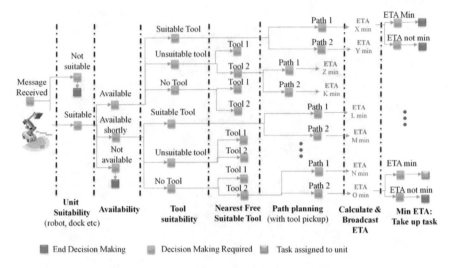

Fig. 4.16 Visualization of the mobile unit consideration in the decision making [79], Reprinted with permission from Pergamon

- *Path planning (with tool pickup if needed)*: the local planner of the mobile unit generates alternative plans for the identified locations and the locations being available for tool pickup.
- *Calculation of the ETA*: for each path generated in the previous step, the Estimated Time for Arrival (ETA) is calculated.
- *Min ETA*: the path with the minimum ETA is selected and considered in the upcoming assignment of tasks for the line's reconfiguration.

The reconfiguration scenarios derived from the control logic, have allowed for the smooth absorption of technical disturbances without disturbing the production flow. The generic design of the control logic can accommodate the control of large-scale applications, involving more mobile units and tools as well as multiple product variants. However, the authors have pointed out limitations that are related to the mobile robots infrastructure e.g. limited examples of mobile units, capable of carrying high payload robotic arms, the mobile robots' power autonomy, automated and collision-free motion planning, etc., as well as for the control and software infrastructure. Namely, there is a need for systematic and automatic generation of the re-configuration alternatives, real-time communication with each resource, through web services, and application of ontologies and semantics technologies for fast data retrieval, combined with reasoning techniques.

Gkournelos et al. (2020) proposed a scalable assembly execution control framework, aiming at facilitating the real reconfigurations for shopfloors, involving mobile manipulators (which consist of arms, torso, as well as a mobile base) [80]. The reconfigurations are realized at two distinct levels (a) process level reconfiguration and (b) resource level and they are coordinated via the execution control framework, as shown

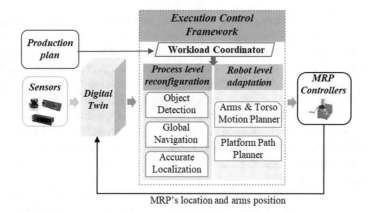

Fig. 4.17 Overall approach [80], Reprinted with permission from Elsevier

in Fig. 4.17. The proposed approach comprises a set of cognition modules for environment and process perception, wrapped as digital services consumed by the execution orchestration mechanism. The implementation of the framework is ROS based and all of the modules are developed with C++ and deployed as ROS nodelets, enabling zero-copy transport communication among them. For the connection with the outside of the framework's shell, the standards interfaces (topics, services, actions) are for the easy and agnostic integration of robot suppliers or digital twin implementations.

A real time updated digital twin of the workplace is responsible for the collection of data and the virtual representation of the shopfloor status. In addition, it enables the use of a unified semantic data model, simplified sensor data sharing, and subsequently simplified control integration and real-time robot behavior adaption. The workload coordinator operates on top of the process and robot reconfiguration levels. It is a simple module, which acts only as a delegator. Its main responsibility is to decode the production plan into robot actions and send these actions for execution. The process level reconfiguration consists of the mechanisms that enable the adaption of the process, based on the shopfloor status at runtime. The following three perception modules allow the online reconfiguration of the process according to the actual process parameters:

- The Object Detection module handles the accurate detection of the position of the parts that are involved in the process. The design of this module enables the easy integration of different detection algorithms, depending on the process or part requirements. Several methods can be used for the detection of parts, such as the contour method detects the object by detecting its outline or else contour, or the Point Pair Feature PPF and Iterative Closest Point ICP algorithms for surface matching which find the spatial relation that aligns two 3D point sets. This module was implemented as a C++ ROS nodelet and it can use any 2D or 3D image published as input for the detection.

- The Global Navigation module deals with the motion of the mobile platform, inside a known shopfloor map, which contains static obstacles. The static obstacles are defined through a mapping process, which takes place before the execution of the production. In case new and unknown obstacles appear, they can be included and avoided through the online information that the digital twin provides. A latter sub-component manages the connection with the digital twin, using ROS topics and services. The Global navigation module is responsible for retrieving from the digital twin, sensor data and semantic information such as the known map of the working area. All these data are fused and a cost map is created with all the static and dynamic objects. This cost map is provided as input to the Platform Path planner module.
- The Accurate Localization module is responsible for covering any accuracy issues that arise during the navigation from one process station to another. The accuracy issues are expected to similarly affect the object detection, due to the limited field of view of the sensors that are commonly used, hence risking the success of the complete workflow. Visual servoing of the mobile platform ensures precise virtual docking with the working tables or fixtures.

The robot level adaptation is the lower level of the Execution Control Framework and it deals with the resource re-organization, meaning that it is responsible for the adaptation of navigation paths and arm motion trajectories. The main goal of this level is to avoid potential collisions with other resources and unmapped obstacles inside the environment. In turn, this requires the coordination of the different controllers, included in the mobile manipulator e.g. mobile base, robot arms, and torso joints. Based on the digital twin geometric data, a planning environment is constructed, containing the information of the robot's current state, the robot kinematics, any objects that are rigidly attached and obstacles to be avoided. The Arms & Torso Motion Planner is dedicated to ensuring safe and collision free motion planning. This module is developed based on MoveIt! The Platform Path Planner is used for online path planning of the mobile platform, aiming at finding a minimum cost plan from a start point to an end point, by considering the static and the dynamic obstacles on the working environment. The path planning component is implemented, based on the ROS navigation stack for mobile robots; thence, the use of the common ROS interfaces (topics, services, actions) is essential. This provides great flexibility in integration of different mobile robots. The vision-based robot control with the output of Accurate localization, a dedicated closed loop PID velocity controller of the mobile, which takes as feedback the position error between the platform and the detected tag, has been developed. Figure 4.18 depicts a sequence diagram, describing the communication and the dataflows among the modules involved.

Karagiannis et al. (2018) suggested a multilayer adaptable control framework for easy configuration and coordination [81]. The authors discussed the design and implementation of a software central execution platform that undergoes the config-uration and control activities, required in a modular production station. XML-based resource configuration and task allocation as well as support for controlling multiple resource types, such as robots, grippers, vision systems, etc. following the "plug &

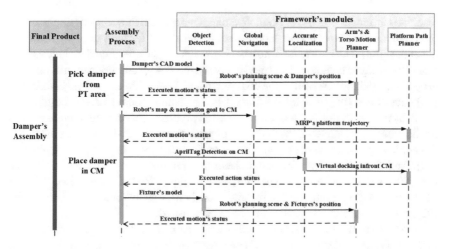

Fig. 4.18 Sequence diagram for task execution [80], Reprinted with permission from Elsevier

play" concept, are the key benefits of the proposed system. The proposed approach is multilayer and can adapt to different assembly cells with the activation of the modules necessary for the execution. The lowest layer comprises the resources and sensors of the assembly cell. The next layer consists of the device drivers and the ROS nodes that are essential for receiving data and forwarding them to the top layer, as well as for the execution of the commands received from the dedicated controllers. The top layer contains the main controller of the framework, which is the core of the proposed software platform and the resource configurator module. These are responsible for executing the program and configuring the existing resources in the cell, based on the info received via an XML file from the programmer. The contribution of this tool is the combination of resource configuration and control modules, allowing for multiple and different resource types and sensors to be integrated and execute a common program flow. Moreover, decisions are taken online, based on the data received from the sensors, by selecting the corresponding task chain that has to be performed in different cases. Furthermore, it can support different configurations and controllers for multiple resources of the same type by increasing the system's flexibility. Finally, it allows the programmer to use parallelism in resource activities by running multiple operations from different resources. Potential sectors could be those of the automotive, aeronautics, and white goods industries.

Maintenance planning and scheduling

Several approaches have been proposed, aiming to reduce or even eliminate unforeseen downtimes, and plan maintenance activities more accurately. The vast availability of field data resulting from ICT advancements, such as the Internet of Things, has triggered the development of new approaches for maintenance, including condition monitoring and predictive maintenance. Carvalho et al. (2019) indicated Random Forests, k-means, and Artificial Neural Networks as the most popular machine learning methods for predictive maintenance, but they also highlighted that the predictive maintenance methodologies presented so far tend to be application-dependent [82]. The multitude of approaches that have been presented include among others the study of data-based and physics-based models for condition monitoring and prognostics [83, 84], as well as the combination of Digital Twins with physics-based [85] data-driven [85] or hybrid [86] models for planning maintenance activities. In more detail, the Digital Twin concept has been proposed for predictive maintenance applications in manufacturing, to enable physics-based predictive maintenance [87, 88] exploiting the knowledge that can derive from the modeling of robot dynamics [89]. An IoT machine learning and orchestrator framework for the detection of failures [90], ARMA modeling and data-driven techniques for fault prediction [91], and Estimation of the Remaining Useful Life, based on Convolutional Neural Networks [92], are some of the approaches that have been tested.

An integrated neural-network-based decision support system, for predictive maintenance of rotational equipment, has been presented [93]. Deep Boltzmann Machines, Deep Belief Networks, and Stacked Auto-Encoders have been employed for the identification of the fault condition of the rolling bearing [94]. A recurrent neural network-based health indicator has been proposed for the RUL prediction of bearings [95]. An ensemble prognostics method considers the effect of degradation on the accuracy of RUL prediction, through the assignment of an optimized, degradation-dependent weight to each learner [96]. Choo et al. (2016) suggested an adaptive Multi-scale Prognostic and Health Management methodology with a hierarchical Markov Decision Process approach to describe and find the optimal policy of a smart manufacturing system [97].

An important issue that comes up, along with the fault diagnosis, is the design of fault features, which usually demands effort and it does not comprise an automated procedure [98]. Shao et al. (2018) tried to get rid of the dependence on manual feature extraction [99]. A Neuro-Genetic algorithm for condition monitoring, fault diagnosis, and evaluation of induction motor, without any additional information [100], as well as a data-driven approach for prognostics, using deep convolution neural networks (DCNN) demanding no prior knowledge of the critical components' degradation process [92], have been proposed. The degradation of the performance of intelligent fault diagnosis has also been addressed by Zhang et al. (2018) using deep learning [101]. Furthermore, decision tree classifiers have been used for the prediction of defective parts in an aluminum injection line [102].

Cloud-based Cyber-Physical Systems, along with the Internet of Things concept, have been proposed for adaptive scheduling and condition-based maintenance [103]. The proposed work presents a cloud-based cyber-physical system, depicted in Fig. 4.19 for adaptive shop-floor scheduling and condition-based maintenance.

The main contribution of the present work is a (a) cost-effective and reliable monitoring system, which integrates data from different sources, by implementing both industrial communication protocols and standards, follows the Industrial IoT and Industry 4.0 paradigms, (b) data analysis algorithms that can easily identify the status of the shop-floor and calculate important key performance indicators in real-time, a monitoring system that includes a wireless sensor network, (c) an adaptive

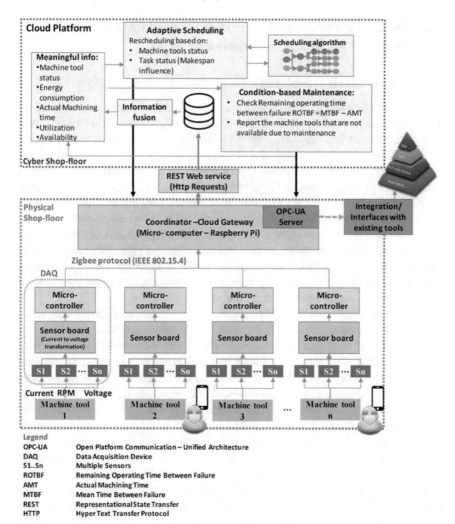

Fig. 4.19 The cloud-based cyber-physical system [103]

scheduling algorithm which consists of a multi-criteria algorithm, capable of taking into consideration, various data from the shop-floor (machines, human operators, etc.) in real-time, as well as input from condition-based maintenance, performing accurate and effective production scheduling and re-scheduling in real-time, (d) the implementation of different modules of the proposed cyber-physical system in a cloud environment, along with technologies for data storage and handling and finally the provision of the different modules (monitoring, scheduling, maintenance) as services upon end-user request.

4.4 Digital Platforms and ICT Technologies for AI Applications in Manufacturing Systems

Availability of data in manufacturing—Big Data and data analytics

Big data are essentially data sets that are so voluminous and complex, rendering the traditional data processing application software inadequate to deal with them. With the increase in data availability and sensor fusion, in the manufacturing industry, the role of data analytics, in the development of smart manufacturing systems, has grown tremendously. For instance, Forbes reported that the global market for marketing analytics is currently around $3.2 billion and is predicted to grow to $6.4 billion by 2026, allowing companies to generate consumer insights and improve strategies to reach them [104]. More and more manufacturers have been applying data and analytics across their processes so as to optimize their supply chains, improve product scheduling and sales forecasting, reduce costs, develop new propositions and monitor machine usage and reliability. The integration of Big Data analysis, in modern manufacturing systems, can significantly accelerate production and development along with identifying quality issues in real time.

The notion of the data analytics comprises a collection of different tool types, namely those based on predictive analytics, data mining, statistics, data mapping, profiling, natural language processing and so on. All these techniques have been around for years. The difference today is that a lot more user organizations are actually using them. That is because most of these techniques adapt well to very large, multi-terabyte data sets with minimal data preparation [105]. Data mining, has been employed for the acquisition of knowledge about the future behavior of a manufacturing system [106]. The adoption of IoT in manufacturing, enables the transition from traditional manufacturing systems to modern digitalized ones. This is realized via data analytics for the production of more value, leading to the generation of new economic opportunities through the re-shaping of industries [107].

Decision making is ubiquitous in the contemporary organizational processes. Big Data analytics is used for the improvement of decision making in different manufacturing stages. The data fusion and the interpretation of data, lead both to the improvement of the product design and to the product use, since the value feedback is collected through sensor-based services, supported by IoT, following collaboration schemes [108]. The method of least squares, in support of vector machines, has been used for the estimation of the manufacturing cost of airframe structural projects [109], back-propagation neural networks and the least squares support vector machines to solve the product life cycle cost estimation [110].

Big data is commonly unstructured and require more real-time analysis. This development calls for new system architectures of data acquisition, transmission, storage, and large-scale data processing mechanisms. Considerations of fault-tolerance, security, and access control are critical in many applications. Both structured and unstructured Product Life Cycle data analytics techniques have been introduced [111]. Kumar et al. (2016) presented a MapReduce framework for automatic pattern recognition in the context of fault diagnosis in cloud-based manufacturing [112].

The scalability and high availability of the IoT and Big Data solutions, without compromising performance requirements, lead to the need for databases that could support the management of massive amounts of data. NoSQL databases are increasingly used in big data and real-time web applications [113]. Following a Column data model, the proposed framework utilizes Cassandra NoSQL database, a leading distributed database [114].

In summary, the increasing use of Data Analytics in modern Manufacturing Systems can lead to an optimized life cycle, lower capital costs and the improvement of process cycle time, resulting in increased yield and faster time to market [115].

Information flow

A significant part of the information flow, required in a manufacturing system, is covered by the Computer Integrated Manufacturing (CIM), which concerns activities related to the manufacturing business. These activities include among others, evaluating and developing different product strategies, analyzing markets and generating forecasts, analyzing product/market characteristics and generating concepts of possible manufacturing systems, designing and analyzing components for machining, inspection, assembly, and all other processes, relating to the nature of the component and/or product (i.e. welding, cutting, etc.), evaluating and/or determining batch sizes, manufacturing capacity, scheduling, and control strategies, relating to the design and fabrication processes involved in the particular product.

The CIM systems consist of subsystems that are integrated into a whole. In turn, the subsystems consist of business planning and support, product design, manufacturing process planning, process control, shop floor monitoring systems, and process automation. The goal of CIM is to use advanced information processing technology, in all areas of the manufacturing industry to:

- Make the total process more productive and efficient.
- Increase product reliability.
- Decrease the cost of production and maintenance, relating both to the manufacturing system as well as to the product.
- Reduce the number of hazardous jobs and increase the involvement of well-educated and competent humans in the manufacturing activity and design.
- Respond to rapid changes in market demand, product modification, and shorter product life cycles.
- Achieve better use of materials, machinery, and personnel, and reduced inventory.
- Achieve better control of production and management of the total manufacturing operation.

Nowadays, the rapidly changing customer demands in combination with the increasing quality and documentation requirements, as well as the complex supply chains, involving stakeholders from all over the world, require flexible production environments, capable of adapting produced goods quickly and efficiently to market demands [116]. In turn, the corresponding information flows are complex and difficult to maintain. Information flows link the operational technology (OT) domain, which generates data and involves industrial and factory automation, supply chain management, asset monitoring, etc., with the information technology (IT) domain, which consumes data and includes automation of business and office, mobile applications, and enterprise web.

The emerging technologies and concepts of Industry 4.0 may fulfill these challenges by enabling end-to-end communication among all production-relevant assets in the production and IT, meaning that modern production environments are equipped with several kinds of Internet of Things (IoT) devices, varying in function and location, collecting huge amounts of data. The main goals of Industry 4.0 include (1) vertical integration of IT systems in production and automation engineering; (2) horizontal integration of various IT systems across the value chain; (3) consistency of engineering over the complete lifecycle; (4) customization of products, through small lots or even lot size one; and (5) new social infrastructures for work.

The major challenge in IT integration is to define the architectures, dictating the information and communication flows that crosscut different Industry 4.0 components. In more detail, interoperability remains a great challenge in smart manufacturing [117], whereas, no consensus has been reached as to the way of addressing compatibility with legacy industrial systems, which often need to be kept operative in the production line [118]. On top of that, security, trust, and privacy remain open research topics [119–121]. Furthermore, a radically new industrial model that will merge the OT and IT domains is necessary so as to ensure the constructive use of the produced data. Storage and processing data management features should be based on data analysis libraries that support the extraction of knowledge, enabled by recent ICT technologies.

A set of reference architectures have been proposed to encompass the major communication challenges that emanate from the difficulty of interfacing different types of computers, purchased by different vendors at various times. Reference architectures serve as blueprints for building and interoperating a software-intensive system and they document the essence from a collection of systems, in a given domain, by providing knowledge on how to develop, standardize, and evolve software systems to this domain. AUTOSAR for the automotive sector [122], ARC-IT for transportation systems [123], and SOA RA for service- oriented systems [124] are some indicative initiatives of this kind, developed by Consortia of major industrial players (such as manufacturers and suppliers) and researchers. Nevertheless, there are a lot of different scenarios and challenges for the implementation of Industry 4.0 systems. Industry 4.0 employs reference architectures with different content, format, and purpose [116]. The implementation of Industry 4.0 communications as well as intelligence, together with big industrial data are, so far, only covered by the IIRA, SITAM, IBM Industry 4.0, and LASFA reference architectures.

Applications of AI techniques are mostly oriented on decision making and control. However, in the age of enhanced connectivity, distributed storage and software execution, big data have enabled AI as embedded in highly interconnected devices that can make them smarter [125]. Artificial intelligence and machine learning methods can enable new decision-making mechanisms by providing insight into production operations, frequently provided as AI services. AI services refer to infrastructures, software, and platforms, provided as services or applications, frequently on the cloud, available off the shelf, and executed on demand. In this way, they reduce the management of complex infrastructures. The backbone of cloud computing is considered to be the Infrastructure as a service (IaaS) that provides access and management of virtual resources, such as servers, storage, operating systems, and networking. In addition, cloud platforms (provided under the Platforms as a Service policy) are service products of cloud applications and can be used within Software as Service (SaaS) architectures, which are cloud applications and adaptive intelligence software [126]. AI and semantic web technologies are discussed in [127], facilitating cross-domain integration of systems and services for smart manufacturing. Case-based reasoning has been applied to the implementation of an intelligent fault detection system for the production o injection molded drippers [128]. The domain expert knowledge is used to determine weights of significant features, which in turn, are used in a local similarity measurement. An artificial immune system approach to case-based reasoning for fault detection diagnosis is discussed in [129].

Digital Twin models have been proposed for the acceleration of the training phase in Machine Learning (ML). For instance, Alexopoulos, Nikolakis, and Chryssolouris (2020) suggested a framework for the implementation of a DT-driven approach to developing ML models, based on an architecture, adapted for the representation of the main entities i.e., the CPS that links the physical world (e.g. machine or robot on the factory floor) to the cyber, through the creation of a digital thread between them. Thus, a Cyber-Physical-Production System is formulated along with the DT that represents the virtual model of the physical system or process. These are linked with the CPS entity, through the data communication channel, being capable of replicating aspects of the behavior of the CPS. Both CPS and DT stacks are defined and implemented, based on the same layered architecture approach.

Context-aware intelligent service systems can be used to provide AI services to people on the shop floor and back office. Such systems may combine key IIoT concepts, such as multi-layered, service-oriented architecture, which integrates several subsystems, e.g. sensor data acquisition together with concepts for developing AI systems that can be combined with the digital twin concept (Fig. 4.20). Industrial Internet of things (IIoT) context-aware information systems [131] can be utilized in the context of decision support for mobile or static operators and supervisors according to their situation. The utilization of context-aware information delivery seems even more relevant in the case of dynamic, semi-structured manufacturing environments in that the workstations are not fixed and change dynamically over. The work has been applied to the shipbuilding industry, where the work areas vary among different projects and the workers are highly mobile and may execute different tasks during their shifts. The implementation of the approach has helped extracting some conclusions which were: 1. thorough understanding of ICT infrastructure and corporate policy restrictions have proven to be important (e.g. cloud solutions not accepted in some cases due to security concerns), 2. installing and deploying Industry 4.0 solutions necessitates considerable effort and commitment from corporate resources due

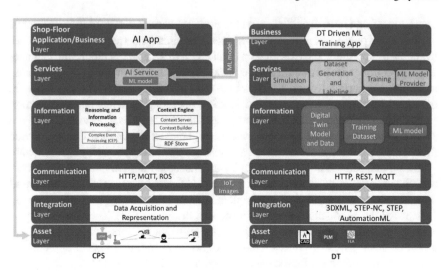

Fig. 4.20 Utilization of the digital twin for the development of ML-based applications for smart manufacturing [130], Reprinted with permission from Taylor & Francis

to the cross-sectorial nature of the solution, 3. successful architectures for manufacturing applications share some common characteristics such as (i) layered architecture approach to ease the management of developing a complex system by isolating independent features and functionalities (Fig. 4.21), (ii) event-driven approach due to the large number of IIoT devices present in manufacturing environments, (iii) context awareness supported by technologies, such as ontologies and semantics to support the dynamic change of user and application context, (iv) finally, it is critical that there is operator/team leader acceptance of the solution, as well as the mentality that such solutions aim to help everyday job activities and support maintaining job positions.

Knowledge representation

An important activity of using AI for manufacturing is knowledge representation. This domain is concerned with the modeling of heterogeneous manufacturing knowledge, the identification of the required level of detail for the modeling, as well as the manner to extract and represent new information using the knowledge. Furthermore, the integration of the knowledge representation with the data scientists' workflow, ensuring that the knowledge representation methodology will be seamlessly accessible, and defining the cost of setting up and maintaining the knowledge representation methodology, are several challenges that are of interest. Knowledge graphs among other tools have been proposed for knowledge representation. Within the AI lifecycle, the knowledge graph is the getaway to all the knowledge about the manufacturing system and processes of an organization, including historical data, the way they are related, their source, the operators involved and the date experiments that have been conducted.

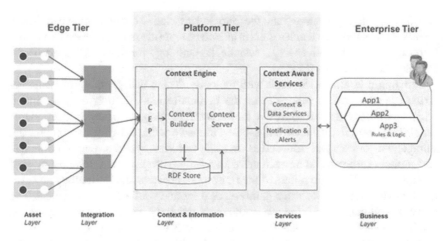

Fig. 4.21 Mapping of the three-tier architecture to the context-aware system layers [131], Reprinted with permission from Taylor & Francis

Two major technologies have usually been employed for knowledge representation; metamodeling and ontologies. A metamodel-based knowledge graph focuses mainly on typing and constraining the ensemble of concepts in the domain. Hence, it facilitates integration into a software engineering process. On the other hand, ontology focuses on common vocabulary, categorization, and the relationships of concepts for large domains. It allows automated reasoning, based on assuming an open world, which allows a better representation of reality.

Knowledge graphs have been proposed for the knowledge representation, in the manufacturing system to make the data analysis process for AI applications more efficient. A knowledge graph gathers all data, information, and knowledge, related to the manufacturing system. It can be structured, based on conceptual models of a knowledge domain, such as the product design and its manufacturing processes. Consequently, domain experts can use real-world concepts, relationships, and vocabulary to describe and solve problems, related to the domain. Knowledge graphs in AI for manufacturing are not yet common ground, which limits the application of AI on the system level, due to the difficulty in finding the relevant data and information. In manufacturing companies, the amount of collected data is huge, as well as the amount of implicit knowledge about the data and the manufacturing processes that hide behind them. Especially in the manufacturing industry, there is a lot of unleveraged domain knowledge, more than in any other (purely data-driven) application domains, the capturing of which would be a strong enabler for the implementation of AI-based applications.

Ontology engineering has been used in multiple fields to provide a common understanding of a domain. In software engineering-related projects, RDF and OWL are two well-established standards, which have been popularized by the semantic web, intending to make the Internet machine-readable by structuring its metadata. DBpedia, a formalization attempt of Wikipedia that allows querying its resources and links to related resources, and the Google Knowledge Graph, which can extract specific information on search results in an info box, are also success stories. In manufacturing, similar initiatives exist, such as the Ontology for Sensors, Observations, Samples, and Actuators (SOSA).

Efthymiou et al. (2015) dealt with the early design and planning of manufacturing systems, and in more detail, the automatic identification of past similar projects, whose reuse will constitute the basis for the design of a new production line [132]. They introduced a knowledge-based framework that the systematic capturing, storage, and retrieval of knowledge on manufacturing systems, permits effective past projects (process and infrastructure knowledge) usage during the early steps of system design. The main pillars of the framework are semantic technology and artificial intelligence approaches, e.g. inference rules and similarity measurement. The semantic technology with the use of the ontology and inference rules supports the detailed and accurate knowledge of the modeling of manufacturing systems. On the other hand, the reasoning with the similarity mechanisms and rules on the semantic data, enhance the identification of new knowledge. In [133], an ontology model for reconfigurable machines is discussed.

Besides ontology and meta-modeling, Fuzzy Petri Nets were also suggested by Liu et al. [134] as a potential modeling technique for knowledge representation and reasoning. A knowledge-based advisory system for multi-material joining is discussed in [135]. The data, including joinable materials, mechanical and design requirements, geometry, etc. were classified and common parameters were stored in a general tree structure DB. A rule-based searching algorithm was used to deliver the knowledge of the joining methods to the structural designers. Wang et al. in [136] proposed a mixed knowledge model combining the features of fuzzy Petri Nets and the learning capability of evolutionary algorithms, using a Genetic Particle Swarm Optimization algorithm to diagnose faults of launch in vehicles, with potential applications in the design of complex products.

Multi-modal context-aware interfaces

A key characteristic of Industry 4.0 is the connection of physical items, such as sensors, devices, and enterprise assets, both to each other and the Internet. In this Internet of Things environment, things can sense more data, become context-aware and provide added-value information to assist people in making decisions. Context-aware information distribution may offer substantial value to manufacturing as it includes task-relevant information, services, or context-driven recommendations, provided to users on a manufacturing shop floor. In this perspective, Alexopoulos et al. (2016) proposed a generic-layered architecture to fit into the manufacturing requirements and serve as a new paradigm to support information distribution and decision-making on the shop floor [137]. Figure 4.22, depicts the information availability in typical legacy systems, which consider only a static pre-defined description of users' roles and a static definition of their context to deliver production-related information. On the contrary, the context and the proposed paradigm account for the definition of roles during the delivery of information, among the factories' legacy/ICT systems, namely the MES and ERP and the people working on the shop floor.

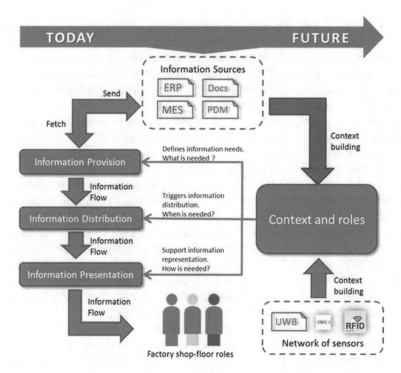

Fig. 4.22 Data delivery in a typical production facilities environment and proposed future information delivery schema that utilises context and roles [137], Reprinted with permission from Taylor & Francis

The proposed paradigm is expected to introduce benefits, directly inherited upon applying information technology to the shop floor, e.g. by reducing or eliminating paper-based information sheets, up-to-date availability of information, related to the product and production, but also reduction in the error-prone manual collection of process data by using sensors in several steps of the processes. Additionally, it is expected to allow the provision of context-based information services to be used for decision support. The right information can be provided to the right people, at the right time on display devices, static or mobile, thanks to the combination of primitive data, captured by sensor devices with production data (available through EPR). Shop-floor production information, captured and analyzed, in real-time, can trigger alerts and notifications. This closes the loop between production planning and execution and potentially results in better quality and process efficiency. Therefore, decision support systems can use the proposed approach as a foundation.

In a similar approach, Belkadi et al. [138] proposed a context-aware knowledge-based system, dedicated to supporting factory agents with the right information at the right time and in the appropriate format, regarding their context of work and level of expertise. Particularly, specific assistance functionalities are dedicated to the workers, in charge of the machine configuration and the realization of manufacturing operations. PGD-based (Proper Generalized Decomposition) algorithms are used for real-time simulation of industrial processes and machine configuration. At a conceptual level, a semantic model is proposed as key enablers for the structuration of the knowledge-based system. Pizoń and Lipski [139] have proposed genetic programming, by providing recommendations, based on multidimensional data sheets for the management of manufacturing processes.

Sotiris Makris et al. (2013) focused on providing instructions to the shop floor workers. In particular, they proposed an algorithm for automatic assembly sequence generation, together with an Augmented Reality (AR) application that visualized the assembly instructions [140]. The algorithm for the generation of the assembly sequence is based on the assumption that the disassembly subtraction paths of the assembly parts are parallel to one or more axes of their local or global axes system. The disassembly sequence and subtraction paths are first generated, based on intersection tests, i.e. movement constraints per part, and then, they are reversed to produce the assembly sequence and insertion paths of the parts. In order for the disassembly sequence to be generated, the 3D objects associated with the assembly parts are placed in a 3D scene and their names in a disassembly array.

The AR component uses data from the AR virtual instructions repository, which in turn, consists of template process tasks. There are several types of screwing actions that are stored in the repository. Depending on the tool (e.g. screwdriver, open wrench, and socket wrench), the instruction templates can be used for the corresponding fasteners (screws, bolts, and nuts). This study has concluded that AR can be used as a final assembly guidance medium, in assistance to both the production engineers and the shop-floor operators, in the generation and usage of assembly instructions, respectively. The sequence generator was used to assist engineers in generating the AR instructions with minimum expert input. The sequence generation and the AR instructions components were connected through a common semantic representation of the sequence and steps. Apart from the cognitive load of both engineers and operators being decreased, the merging of these technologies can significantly reduce the time between product design and production, especially in the systems that handle customized products and require higher flexibility. These technologies could, in the future, replace traditional methods for the manual creation of paper-based assembly instructions, which will be boosted by the commercialization of new AR equipment.

A semantic-based Augmented Reality application, extracting information from CAD/PDM systems for real-time operator support, has been demonstrated in an automotive case study, using a computer tablet [141]. In a work with a similar objective, Zhang, Ong, and Nee (2011) proposed inertial sensors and infrared enhanced computer vision for assembly guidance in augmented reality [142]. Another aspect of interest is the feedback about the operators' performance. In this scope, there is a feature-based assembly recognition algorithm for the evaluation of human performance, in assembly operations as proposed [143]. AR-based applications have also been proposed to support design and assembly in tractor [144] and automotive [145] companies.

Makris et al. (2016) proposed an AR application by providing production and process information to enrich the operator's immersion in safety mechanisms, in

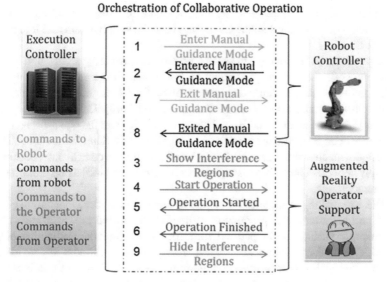

Fig. 4.23 Execution of collaborative operation sequence [146], Reprinted with permission from Edition Colibri AG

support of hybrid manufacturing [146]. In more detail, the design and implementation of augmented reality (AR) tool, in aid of operators being in a hybrid environment, there was a human and robot collaborative industrial environment presented. The system is responsible not only for the provision of production and process-related information, but also for enhancing the operators' immersion in safety mechanisms that emanate from the collaborative workspace (Fig. 4.23). A service-based station controller, responsible for orchestrating the flow of information to the operator, has enabled the integration of the developed system according to the task execution status.

Supply chains

Every firm is a collection of activities that are performed to design, produce, market, deliver, and support its product. A firm's value chain and the way it performs individual activities reflect its history, its strategy, its approach to implementing its strategy, and the underlying economics of the activities themselves. The supply chain is a special case of the value chain of those companies that manufacture or distribute physical products. The value chain has also been called the value-added chain to focus on the firm's ultimate objective of adding value to its products or services, at each stage of its chain. A company's supply chain comprises geographically dispersed facilities, where raw materials, intermediate products, or finished products are acquired, transformed, stored, or sold, and transportation links that connect facilities along which products flow. The facilities may be operated by the company, or they may be operated by vendors, customers, third-party providers, or other firms. The company's goal is to add value to its products as they pass through its supply chain and to transport them to geographically dispersed markets in the correct quantities, with the correct specifications, at the correct time, and at a competitive cost. Therefore, the next paragraphs discuss how the supply chain and logistics are managed.

AI-powered technologies e.g. pattern recognition, expert systems, and artificial neural networks help manage complexity [147], and in the case of supply chain, design and management can reduce waste, allow for lower process cycle time, and enable real-time monitoring and error-free production [148]. The Chaos theory-based analysis [149], the Branch and Bound algorithms [150] and the agent-based models [151] have been proposed for the design of supply chains. Saint Germain et al. [152] suggested a mechanism, based on PROSA and ants for the control and coordination of supply networks. Inventory management is discussed in [153] using reinforcement learning methods.

Decision-making in logistics is a cognitive stressful task, since logistics in general, is a complex system. The production of highly customized products, calls for the dynamic querying of supply partners for information about the availability of parts [154]. A method of dynamically querying supply chain partners to provide real-time or near real-time information, regarding the availability of parts is required for the production of highly customizable products. This method utilizes Internet-based communication and real-time information from RFID sensors. The control logic handles the customization orders adequately by enabling the supply chain to adapt to market variation, thus reducing significantly the order to delivery time. The feasibility of this approach is demonstrated with its implementation in a typical automotive case.

Internet-based communication and specially designed frameworks can enable the integration of heterogeneous information systems, while the performance of the supply chain can be improved through hierarchical modeling [155]. In current practice, companies in the manufacturing industry operate globally in order to expand the limits of their business and integrate their operations with those of their business partners. The growth of the Internet and the software technologies arising from it provide the means for this globalization. In this work, it is demonstrated how modern information technology can support the communication of different partners and enable the information flow within the value-added chain. Moreover, it is described how the efficiency of a supply chain can be improved with the application of a generic hierarchical model through the proper planning of critical manufacturing operations. Supply chain management is a relatively new term, crystallizing concepts about integrated business planning and it was suggested by the academic community in the 1950s. This work is based on a 'real-life' ship repair scenario, where it is assumed that a ship has to visit a shipyard for planned maintenance. It is a typical situation that at least two partners will participate in the process: the shipyard and the ship owner. The business process is as follows (Fig. 4.24):

The system uses event-driven simulation to simulate the operation of the shipyard and the execution of the workload from the shipyard's resources. The simulation mechanism releases the workload to the Job Shops and Work Centers, respecting the user-defined precedence relationships. In each Work Centre, an assignment mechanism decides which Task will be assigned to the available Resources. The system

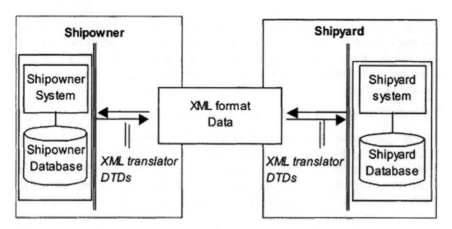

Fig. 4.24 The value-added chain communication model [155], Reprinted with permission from Taylor & Francis

simulates the operation of the production facilities, either for a certain period (user-specified) or until all the Tasks have been processed by the Resources. In either case, a detailed schedule for each Resource is produced in a graphic or an alphanumeric format.

Modern information technology has been used in order to support the Value-Added Chain in the maritime industry. A software mechanism was developed to make the communication of the interrelated partners efficient. It is demonstrated in this study that the adoption of such an approach is feasible and the communication efforts are reduced. The use of XML for the implementation of communication in the value-added chain offers advantages, due to its simplicity and openness.

Additionally, the planning method applied for the coordination of the ship repair operations produces adequate and easy-to-use results. This method requires the modeling of the ship repair facilities and the workload with the help of a hierarchical model. Using the implemented approach, the user can select an appropriate dispatching rule or the multi-criteria decision-making method, for the production of a suitable plan.

Decision-making support for the design and operation of manufacturing networks [156]. The presented method and tool can support strategic level decisions, related to the design of efficient manufacturing network configurations. This work focuses on the design and operation of manufacturing networks, based on a multi-objective decision-making and simulation approach. The alternative network designs are evaluated through a set of multiple conflicting criteria, including dynamic complexity, reliability, cost, time, quality, and environmental footprint. Moreover, the impact of demand volatility on the operational performance of these networks is investigated through simulation. Decentralized Manufacturing Network (DMN) and Centralized (CMN) configurations have been modeled and their performance is compared. Two methodologies are used in the decision-making process for the generation and evaluation of manufacturing network alternatives, namely those of the Exhaustive Search

Algorithm (EXS) and the Intelligent Search Algorithm (ISA). The incorporation of criteria of cost, time, environmental impact, and quality have encapsulated some of the most significant objectives that manufacturing industries are striving to achieve nowadays. Moreover, the inclusion of complexity, as a decision-making criterion, has depicted the operational characteristics of the networks.

Forecast of demand for dispatching—clustering to deal with design problems

A critical input to the manufacturing system that highly impacts its operation and even determines the system's sustainability is the customers' demand. The variation that is frequently observed in customers' demand is accompanied by several challenges, in the design of production, related to the fast and inexpensive customization of the goods. The clustering technique has been proposed to address this problem. Utilizing the Design Structure Matrix (DSM) as a tool for the representation of the interactions of a system's elements, several clustering algorithms have been developed based on GAs [157].

The customer's behavior has been modeled via the Bayesian networks [158]. To deal with a customer's individualized demand, a hybrid MPA-GSO-DNN model is suggested in [159] for personalized recommendation of service composition, regarding a manufacturing group with shared manufacturing services. Customers can independently define personalized manufacturing customization, task release, manufacturing configuration, and functions. The application of a personalized recommendation can greatly improve the customers' efficiency and satisfaction.

In [160] a probabilistic inference method for the quantification of a buyer's likelihood to purchase a highly customized product was investigated. This method was based on the principles of Bayesian networks, and it was integrated into an internet-based application, where the supply chain partners could provide real-time or near real-time information. The supply chain model utilizes the Internet to coordinate the entire supply chain by adapting it to market demand. Moreover, the Bayesian networks-based model considers a set of values of critical parameters and enables the quantification of the probability of a customer's likely response, under certain delivery dates for her/his vehicle, additional cost for the OEM, the parts' availability time in the supply chain and the suppliers' capacity. The proposed supply chain control logic has produced robust plans by ensuring the supply of the right part at the right time, whereas at reasonable costs. Quantifying the likelihood that a customer will proceed to buy the final product, has allowed a consistent production schedule to be maintained, avoiding any last minute changes, which cause quality defects, as demonstrated in an automotive case study.

The clustering of a product's components into modules is an effective means of creating modular architectures that are more efficient and effective to be reconfigured into producing personalized products. Pandremenos and Chryssolouris (2011) investigated the clustering efficiency with the interactions of a product's components, and interesting observations were extracted [161]. A novel clustering method utilizing neural network algorithms and design structure matrices has been introduced. The method can reorganize the components of a product in clusters, in order for the interactions to be maximized inside and minimized outside the clusters. In addition, a multi-criteria decision-making approach was proposed to efficiently identify the different clustering alternatives, having derived from the network, to be evaluated. Finally, a case study is presented to demonstrate and assess the method's application. The derived algorithmic clustering has proven to be more efficient, compared to the empirical one, and thus, it can be used by design engineers as an effective tool for the derivation of product clustering alternatives.

A car's Body-in-White has been selected as a case study for the proposed algorithm to be applied and evaluated. The Body-in-White design structure matrix consists of 38 parts and 108 interactions among them. In order to assess the clustering efficiency, the modularity performance was calculated to have a value of 0.89. Since this value reveals a modular architecture, the clustering is not expected to be very efficient.

Similarities in the front and the rear-end modules are observed between the clustering of the proposed method and the empirical clustering. The evaluation of the two approaches (Figs. 4.25 and 4.26) has shown a more efficient clustering in the case of the algorithm. Therefore, given that the feasibility of the clustering architectures and the integrity of the modules during assembly should be investigated, this approach, was suggested for use by design engineers as a tool for the derivation of product clustering suggestions, during the new product family/platform design.

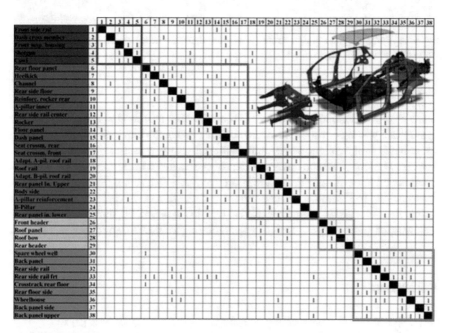

Fig. 4.25 Body-in-White design structure matrix empirical clustering [161], Reprinted with permission from Taylor & Francis

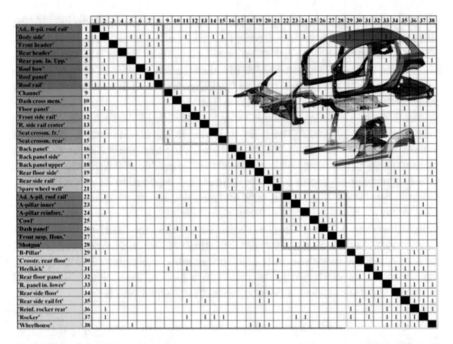

Fig. 4.26 Body-in-White design structure matricx algorithm-based clustering [161], Reprinted with permission from Taylor & Francis

Financial estimations

Aside from the demand, the estimation of the cost unit is also important for the appropriate pricing of products. A method for a product's unit cost estimation, using agent-based fuzzy collaborative intelligence with entropy as a consensus measure, is discussed in [162]. After a consensus has been reached a back-propagation network is employed to de-fuzzify the result. Moreover, an SVM with a genetic algorithm is described in [163] for accurately predictive fabrication costs via learning and curve fitting. Real-world data have been used to enable the described knowledge discovery mechanisms, predicting the costs of manufacturing TFT-LCD fabrication equipment.

References

1. Chryssolouris, G.: Manufacturing Systems: Theory and Practice. Springer (2006)
2. Rekiek, B., Dolgui, A., Delchambre, A., Bratcu, A.: State of art of optimization methods for assembly line design. Annu. Rev. Control. **26 II**, 163–174 (2002). https://doi.org/10.1016/S1367-5788(02)00027-5
3. Sahu, A., Pradhan, S.K.: Quantitative analysis and optimization of production line based on multiple evaluation criteria using discrete event simulation: a review. In: International Conference on Automatic Control and Dynamic Optimization Techniques, ICACDOT, pp. 612–617. Institute of Electrical and Electronics Engineers Inc (2017)

4. Hu, S.J., Ko, J., Weyand, L., Elmaraghy, H.A., Lien, T.K., Koren, Y., Bley, H., Chryssolouris, G., Nasr, N., Shpitalni, M.: Assembly system design and operations for product variety. CIRP Ann. Manuf. Technol. **60**, 715–733 (2011). https://doi.org/10.1016/j.cirp.2011.05.004
5. Spieckermann, S., Gutenschwager, K., Heinzel, H., Voß, S.: Simulation-based optimization in the automotive industry—a case study on body shop design. Simulation **75**, 276–286 (2000)
6. Hsieh, S.J.: Hybrid analytic and simulation models for assembly line design and production planning. Simul. Model. Pract. Theory. **10**, 87–108 (2002). https://doi.org/10.1016/S1569-190X(02)00063-1
7. Moriz, N., Maier, A., Niggemann, O.: AutomationML as a Basis for Offline- and Realtime-Simulation. In: Automation ML as a Basis for Offline—And Realtime-Simulation (2011)
8. Çakar, T., Cil, I.: Artificial neural networks for design of manufacturing systems and selection of priority rules. Int. J. Comput. Integr. Manuf. **17**, 195–211 (2004). https://doi.org/10.1080/09511920310001607078
9. Chryssolouris, G., Lee, M., Pierce, J., Domroese, M.: Use of neural networks for the design of manufacturing systems. Manuf. Rev. (Les Ulis) (1990)
10. Izui, K., Murakumo, Y., Suemitsu, I., Nishiwaki, S., Noda, A., Nagatani, T.: Multiobjective layout optimization of robotic cellular manufacturing systems. Comput. Ind. Eng. **64**, 537–544 (2013). https://doi.org/10.1016/j.cie.2012.12.003
11. Baykasoğlu, A.: Gene expression programming based meta-modelling approach to production line design. Int. J. Comput. Integr. Manuf. **21**, 657–665 (2008). https://doi.org/10.1080/09511920701370753
12. Rekiek, B., Delchambre, A.: Assembly line design: the balancing of mixed-model hybrid assembly lines with genetic algorithms. Springer (2006)
13. Jiang, S., Nee, A.Y.C.: A novel facility layout planning and optimization methodology. CIRP Ann. Manuf. Technol. **62**, 483–486 (2013). https://doi.org/10.1016/j.cirp.2013.03.133
14. Ho, Y.C., Moodie, C.L.: A heuristic operation sequence-pattern identification method and its applications in the design of a cellular flexible assembly system. Int. J. Comput. Integr. Manuf. **7**, 163–174 (1994). https://doi.org/10.1080/09511929408944606
15. Liu, Q., Meller, R.D.: A sequence-pair representation and MIP-model-based heuristic for the facility layout problem with rectangular departments. IIE Trans. **39**, 377–394 (2007). https://doi.org/10.1080/07408170600844108
16. Achanga, P., Shehab, E., Roy, R., Nelder, G.: A fuzzy-logic advisory system for lean manufacturing within SMEs. Int. J. Comput. Integr. Manuf. **25**, 839–852 (2012). https://doi.org/10.1080/0951192X.2012.665180
17. Ayağ, Z., Özdemir, R.G.: A combined fuzzy AHP-goal programming approach to assembly-line selection. J. Intell. Fuzzy Syst. **18**, 345–362 (2007)
18. Sysoev, V., Dolgui, A.: A Pareto optimization approach for manufacturing system design. In: Proceedings of the International Conference on Industrial Engineering and Production Management, pp. 116–125 (1999)
19. Alexopoulos, K., Mourtzis, D., Papakostas, N., Chryssolouris, G.: DESYMA: assessing flexibility for the lifecycle of manufacturing systems **45**, 1683–1694 (2007). https://doi.org/10.1080/00207540600733501
20. Michalos, G., Fysikopoulos, A., Makris, S., Mourtzis, D., Chryssolouris, G.: Multi criteria assembly line design and configuration—an automotive case study. CIRP J. Manuf. Sci. Technol. **9**, 69–87 (2015). https://doi.org/10.1016/j.cirpj.2015.01.002
21. Michalos, G., Makris, S., Mourtzis, D.: An intelligent search algorithm-based method to derive assembly line design alternatives. Int. J. Comput. Integr. Manuf. **25**, 211–229 (2012). https://doi.org/10.1080/0951192X.2011.627949
22. Subramaniyan, M., Skoogh, A., Salomonsson, H., Bangalore, P., Bokrantz, J.: A data-driven algorithm to predict throughput bottlenecks in a production system based on active periods of the machines. Comput. Ind. Eng. **125**, 533–544 (2018). https://doi.org/10.1016/J.CIE.2018.04.024
23. Noktehdan, A., Karimi, B., Husseinzadeh Kashan, A.: A differential evolution algorithm for the manufacturing cell formation problem using group based operators. Expert Syst. Appl. **37**, 4822–4829 (2010). https://doi.org/10.1016/J.ESWA.2009.12.033

24. Arkouli, Z., Michalos, G., Makris, S.: On the selection of ergonomics evaluation methods for human centric manufacturing tasks. Procedia CIRP. **107**, 89–94 (2022). https://doi.org/10.1016/J.PROCIR.2022.04.015

25. Battini, D., Calzavara, M., Otto, A., Sgarbossa, F.: Preventing ergonomic risks with integrated planning on assembly line balancing and parts feeding. Int. J. Prod. Res. **55**, 7452–7472 (2017). https://doi.org/10.1080/00207543.2017.1363427

26. Alexopoulos, K., Mavrikios, D., Chryssolouris, G.: ErgoToolkit: an ergonomic analysis tool in a virtual manufacturing environment. Int. J. Comput. Integr. Manuf. **26**, 440–452 (2013). https://doi.org/10.1080/0951192X.2012.731610

27. Kousi, N., Michalos, G., Makris, S., Chryssolouris, G.: Short—term planning for part supply in assembly lines using mobile robots. In: Procedia CIRP, pp. 371–376. Elsevier (2016)

28. Chawla, V.K., Chanda, A.K., Angra, S.: A clonal selection algorithm for minimizing distance travel and back tracking of automatic guided vehicles in flexible manufacturing system. J. Inst. Eng. (India): Ser. C **100**(3), 401–410 (2018). https://doi.org/10.1007/S40032-018-0447-5

29. Müller, C., Grunewald, M., Spengler, T.S.: Redundant configuration of robotic assembly lines with stochastic failures. Int. J. Prod. Res. **56**, 3662–3682 (2018). https://doi.org/10.1080/00207543.2017.1406672

30. Kulak, O.: A decision support system for fuzzy multi-attribute selection of material handling equipments. Expert Syst. Appl. **29**, 310–319 (2005). https://doi.org/10.1016/J.ESWA.2005.04.004

31. Alexopoulos, K., Papakostas, N., Mourtzis, D., Chryssolouris, G.: A method for comparing flexibility performance for the lifecycle of manufacturing systems under capacity planning constraints **49**, 3307–3317 (2010). https://doi.org/10.1080/00207543.2010.482566

32. Alexopoulos, K., Koukas, S., Boli, N., Mourtzis, D.: Resource planning for the installation of industrial product service systems. In: IFIP Advances in Information and Communication Technology, pp. 205–213. Springer New York LLC (2017)

33. Chryssolouris, G., Dicke, K., Lee, M.: An approach to real-time flexible scheduling. Int. J. Flex. Manuf. Syst. **6**, 235–253 (1994). https://doi.org/10.1007/BF01328813

34. Kousi, N., Koukas, S., Michalos, G., Makris, S., Chryssolouris, G.: Service oriented architecture for dynamic scheduling of mobile robots for material supply. In: Procedia CIRP, pp. 18–22. Elsevier B.V. (2016)

35. Giordani, S., Lujak, M., Martinelli, F.: A distributed multi-agent production planning and scheduling framework for mobile robots. Comput. Ind. Eng. **64**, 19–30 (2013). https://doi.org/10.1016/j.cie.2012.09.004

36. Wilcox, R., Shah, J.: Optimization of multi-agent workflow for human-robot collaboration in assembly manufacturing. In: Infotech@Aerospace 2012. American Institute of Aeronautics and Astronautics, Reston, Virigina (2012)

37. Chen, F., Sekiyama, K., Huang, J., Sun, B., Sasaki, H., Fukuda, T.: An assembly strategy scheduling method for human and robot coordinated cell manufacturing. Int. J. Intell. Comput. Cybern. **4**, 487–510 (2011). https://doi.org/10.1108/17563781111186761

38. Hasgül, S., Saricicek, I., Ozkan, M., Parlaktuna, O.: Project-oriented task scheduling for mobile robot team. J. Intell. Manuf. **20**, 151–158 (2009). https://doi.org/10.1007/s10845-008-0228-8

39. Chryssolouris, G., Dicke, K., Lee, M.: An approach to short interval scheduling for discrete parts manufacturing. Int. J. Comput. Integr. Manuf. **4**, 157–168 (1991). https://doi.org/10.1080/09511929109440491

40. Chryssolouris, G., Dicke, K., Lee, M.: On the resources allocation problem. Int. J. Prod. Res. **30**, 2773–2795 (1992). https://doi.org/10.1080/00207549208948190

41. Chryssolouris, G., Pierce, J., Dicke, K.: An approach for allocating manufacturing resources to production tasks. J. Manuf. Syst. **10**, 368–382 (1991). https://doi.org/10.1016/0278-6125(91)90055-7

42. Kolisch, R., Drexl, A.: Adaptive search for solving hard project scheduling problems. Nav. Res. Logist. **43**, 23–40 (1996). https://doi.org/10.1002/(SICI)1520-6750(199602)43:1%3c23::AID-NAV2%3e3.0.CO;2-P

43. Li, K.Y., Willis, R.J.: An iterative scheduling technique for resource-constrained project scheduling. Eur. J. Oper. Res. **56**, 370–379 (1992). https://doi.org/10.1016/0377-2217(92)903 20-9

44. Chryssolouris, G., Subramaniam, V.: Dynamic scheduling of manufacturing job shops using extreme value theory. Prod. Plan. Control. **11**, 122–132 (2000). https://doi.org/10.1080/095 372800232324

45. Chryssolouris, G., Lee, M., Domroese, M.: The use of neural networks in determining operational policies for manufacturing systems. J. Manuf. Syst. **10**, 166–175 (1991). https://doi.org/10.1016/0278-6125(91)90018-W

46. Chryssolouris, G., Subramaniam, V.: Dynamic scheduling of manufacturing job shops using genetic algorithms. J. Intell. Manuf. **12**, 281–293 (2001). https://doi.org/10.1023/A:101125 3011638

47. Kádár, B., Lengyel, A., Monostori, L., Suginishi, Y., Pfeiffer, A., Nonaka, Y.: Enhanced control of complex production structures by tight coupling of the digital and the physical worlds. CIRP Ann. Manuf. Technol. **59**, 437–440 (2010). https://doi.org/10.1016/j.cirp.2010. 03.123

48. Raj, J.A., Ravindran, D., Saravanan, M., Prabaharan, T.: Simultaneous scheduling of machines and tools in multimachine flexible manufacturing systems using artificial immune system algorithm. Int. J. Comput. Integr. Manuf. **27**, 401–414 (2014). https://doi.org/10.1080/095 1192X.2013.834461

49. Luo, X., Li, W., Tu, Y., Xue, D., Tang, J.: Optimal resource allocation for hybrid flow shop in one-of-a-kind production. Int. J. Comput. Integr. Manuf. **23**, 146–154 (2010). https://doi.org/10.1080/09511920903440339

50. Tuncel, E., Zeid, A., Kamarthi, S.: Solving large scale disassembly line balancing problem with uncertainty using reinforcement learning. J. Intell. Manuf. **25**, 647–659 (2014). https://doi.org/10.1007/s10845-012-0711-0

51. Shahrabi, J., Adibi, M.A., Mahootchi, M.: A reinforcement learning approach to parameter estimation in dynamic job shop scheduling. Comput. Ind. Eng. **110**, 75–82 (2017). https://doi.org/10.1016/j.cie.2017.05.026

52. Zhang, Y., Wang, J., Liu, S., Qian, C.: Game theory based real-time shop floor scheduling strategy and method for cloud manufacturing. Int. J. Intell. Syst. **32**, 437–463 (2017). https://doi.org/10.1002/int.21868

53. Lalas, C., Mourtzis, D., Papakostas, N., Chryssolouris, G.: A simulation-based hybrid backwards scheduling framework for manufacturing systems. Int. J. Comput. Integr. Manuf. **19**, 762–774 (2006). https://doi.org/10.1080/09511920600678827

54. Chryssolouris, G., Papakostas, N., Mourtzis, D.: A decision-making approach for nesting scheduling: a textile case. Int. J. Prod. Res. **38**, 4555–4564 (2000). https://doi.org/10.1080/00207540050205299

55. Chryssolouris, G.: MADEMA: an approach to intelligent manufacturing systems. CIM Rev. **3**, 11–17 (1987)

56. Mourtzis, D., Papakostas, N., Chryssolouris, G.: An approach to planning of textile manufacturing operations: a scheduling method. In: Proceedings of the IFIP WG5, pp. 131–145 (1995)

57. Mourtzis, D., Doukas, M., Vlachou, E.: A mobile application for knowledge-enriched short-term scheduling of complex products. Logist. Res. **9**, 1–17 (2016). https://doi.org/10.1007/s12159-015-0130-7

58. Giannelos, N., Papakostas, N., Mourtzis, D., Chryssolouris, G.: Dispatching policy for manufacturing jobs and time-delay plots. Int. J. Comput. Integr. Manuf. **20**, 329–337 (2007). https://doi.org/10.1080/09511920600786604

59. Michalos, G., Makris, S., Rentzos, L., Chryssolouris, G.: Dynamic job rotation for workload balancing in human based assembly systems. CIRP J. Manuf. Sci. Technol. **2**, 153–160 (2010). https://doi.org/10.1016/j.cirpj.2010.03.009

60. Hao, G., Lai, K.K., Tan, M.: A neural network application in personnel scheduling. Ann. Oper. Res. **128**, 65–90 (2004). https://doi.org/10.1023/B:ANOR.0000019099.29005.17

61. Sabar, M., Montreuil, B., Frayret, J.-M.: Competency and preference based personnel scheduling in large assembly lines. Int. J. Comput. Integr. Manuf. **21**, 468–479 (2008). https://doi.org/10.1080/09511920701574842
62. Zülch, G., Rottinger, S., Vollstedt, T.: A simulation approach for planning and re-assigning of personnel in manufacturing. Int. J. Prod. Econ. **90**, 265–277 (2004). https://doi.org/10.1016/j.ijpe.2003.11.008
63. Techawiboonwong, A., Yenradee, P., Das, S.K.: A master scheduling model with skilled and unskilled temporary workers. Int. J. Prod. Econ. **103**, 798–809 (2006). https://doi.org/10.1016/j.ijpe.2005.11.009
64. Michalos, G., Makris, S., Mourtzis, D.: A web based tool for dynamic job rotation scheduling using multiple criteria. CIRP Ann. Manuf. Technol. **60**, 453–456 (2011). https://doi.org/10.1016/j.cirp.2011.03.037
65. Efthymiou, K., Pagoropoulos, A., Mourtzis, D.: Intelligent scheduling for manufacturing systems: A case study. In: Lecture Notes in Mechanical Engineering, pp. 1153–1164. Springer Heidelberg (2013)
66. Kousi, N., Koukas, S., Michalos, G., Makris, S.: Scheduling of smart intra–factory material supply operations using mobile robots. Int. J. Prod. Res. **57**, 801–814 (2019). https://doi.org/10.1080/00207543.2018.1483587
67. Siatras, V., Nikos, N., Kosmas, A., Dimitris, M.: A toolbox of agents for scheduling the paint shop in bicycle industry. Procedia CIRP. **107**, 1156–1161 (2022). https://doi.org/10.1016/J.PROCIR.2022.05.124
68. Cho, S., Prabhu, V.V.: Distributed adaptive control of production scheduling and machine capacity. J. Manuf. Syst. **26**, 65–74 (2007). https://doi.org/10.1016/j.jmsy.2007.10.002
69. Saaty, T.L.: Decision making with the analytic hierarchy process. Int. J. Serv. Sci. **1**, 83–98 (2008)
70. Monostori, L., Kádár, B., Pfeiffer, A., Karnok, D.: Solution approaches to real-time control of customized mass production. CIRP Ann. Manuf. Technol. **56**, 431–434 (2007). https://doi.org/10.1016/j.cirp.2007.05.103
71. Yan, H. Sen, Yang, H.B., Dong, H.: Control of knowledgeable manufacturing cell with an unreliable agent. J. Intell. Manuf. **20**, 671–682 (2009). https://doi.org/10.1007/s10845-008-0156-7
72. Mařík, V., McFarlane, D.: Industrial adoption of agent-based technologies (2005)
73. Tang, H.P., Wong, T.N.: Reactive multi-agent system for assembly cell control. Robot. Comput. Integr. Manuf. **21**, 87–98 (2005). https://doi.org/10.1016/j.rcim.2004.04.001
74. Zattar, I.C., Ferreira, J.C.E., Rodrigues, J.G.G.G., de Sousa, C.H.B.: A multi-agent system for the integration of process planning and scheduling using operation-based time-extended negotiation protocols. Int. J. Comput. Integr. Manuf. **23**, 441–452 (2010). https://doi.org/10.1080/09511921003665775
75. Mahesh, M., Ong, S.K., Nee, A.Y.C.: A web-based multi-agent system for distributed digital manufacturing. Int. J. Comput. Integr. Manuf. **20**, 11–27 (2007). https://doi.org/10.1080/09511920600710927
76. Bayar, N., Darmoul, S., Hajri-Gabouj, S., Pierreval, H.: Using immune designed ontologies to monitor disruptions in manufacturing systems. Comput. Ind. **81**, 67–81 (2016). https://doi.org/10.1016/J.COMPIND.2015.09.004
77. Huang, G.Q., Zhang, Y.F., Jiang, P.Y.: RFID-based wireless manufacturing for walking-worker assembly islands with fixed-position layouts. Robot. Comput. Integr. Manuf. **23**, 469–477 (2007). https://doi.org/10.1016/j.rcim.2006.05.006
78. Torres, P.M.B.: Text recognition for objects identification in the industry. Lecture Notes in Networks and Systems. **20**, 126–131 (2017). https://doi.org/10.1007/978-3-319-63091-5_15
79. Michalos, G., Sipsas, P., Makris, S., Chryssolouris, G.: Decision making logic for flexible assembly lines reconfiguration. Robot. Comput. Integr. Manuf. **37**, 233–250 (2016). https://doi.org/10.1016/j.rcim.2015.04.006
80. Gkournelos, C., Kousi, N., Bavelos, A.C., Aivaliotis, S., Giannoulis, C., Michalos, G., Makris, S.: Model based reconfiguration of flexible production systems. In: Procedia CIRP, pp. 80–85. Elsevier B.V. (2020)

81. Karagiannis, P., Giannoulis, C., Michalos, G., Makris, S.: Configuration and control approach for flexible production stations. In: Procedia CIRP. pp. 166–171. Elsevier B.V. (2018)

82. Carvalho, T.P., Soares, F.A.A.M.N., Vita, R., Francisco, R. da P., Basto, J.P., Alcalá, S.G.S.: A systematic literature review of machine learning methods applied to predictive maintenance. Comput. Ind. Eng. **137**, 106024 (2019). https://doi.org/10.1016/j.cie.2019.106024

83. Peng, Y., Dong, M., Zuo, M.J.: Current status of machine prognostics in condition-based maintenance: a review. Int. J. Adv. Manuf. Technol. **50**, 297–313 (2010). https://doi.org/10.1007/s00170-009-2482-0

84. Jardine, A.K.S., Lin, D., Banjevic, D.: A review on machinery diagnostics and prognostics implementing condition-based maintenance. Mech. Syst. Signal Process. **20**, 1483–1510 (2006). https://doi.org/10.1016/J.YMSSP.2005.09.012

85. Sikorska, J.Z., Hodkiewicz, M., Ma, L.: Prognostic modelling options for remaining useful life estimation by industry. Mech. Syst. Signal Process. **25**, 1803–1836 (2011). https://doi.org/10.1016/j.ymssp.2010.11.018

86. Aivaliotis, P., Arkouli, Z., Georgoulias, K., Makris, S.: Degradation curves integration in physics-based models: towards the predictive maintenance of industrial robots. Robot. Comput. Integr. Manuf. **71**, 102177 (2021). https://doi.org/10.1016/j.rcim.2021.102177

87. Aivaliotis, P., Georgoulias, K., Chryssolouris, G.: The use of Digital Twin for predictive maintenance in manufacturing. Int. J. Comput. Integr. Manuf. **32**, 1067–1080 (2019). https://doi.org/10.1080/0951192X.2019.1686173

88. Aivaliotis, P., Georgoulias, K., Arkouli, Z., Makris, S.: Methodology for enabling Digital Twin using advanced physics-based modelling in predictive maintenance. Procedia CIRP. **81**, 417–422 (2019). https://doi.org/10.1016/j.procir.2019.03.072

89. Arkouli, Z., Aivaliotis, P., Makris, S.: Towards accurate robot modelling of flexible robotic manipulators. In: Procedia CIRP, pp. 497–501. Elsevier B.V. (2020)

90. Carvajal Soto, J.A., Tavakolizadeh, F., Gyulai, D.: An online machine learning framework for early detection of product failures in an Industry 4.0 context. Int. J. Comput. Integr. Manuf. **32**, 452–465 (2019). https://doi.org/10.1080/0951192X.2019.1571238

91. Baptista, M., Sankararaman, S., de Medeiros, I.P., Nascimento, C., Prendinger, H., Henriques, E.M.P.: Forecasting fault events for predictive maintenance using data-driven techniques and ARMA modeling. Comput. Ind. Eng. **115**, 41–53 (2018). https://doi.org/10.1016/j.cie.2017.10.033

92. Li, X., Ding, Q., Sun, J.Q.: Remaining useful life estimation in prognostics using deep convolution neural networks. Reliab. Eng. Syst. Saf. **172**, 1–11 (2018). https://doi.org/10.1016/j.ress.2017.11.021

93. Wu, S.J., Gebraeel, N., Lawley, M.A., Yih, Y.: A neural network integrated decision support system for condition-based optimal predictive maintenance policy. IEEE Trans. Syst. Man Cybern. Part A Syst. Hum. **37**, 226–236 (2007). https://doi.org/10.1109/TSMCA.2006.886368

94. Chen, Z., Deng, S., Chen, X., Li, C., Sanchez, R.V., Qin, H.: Deep neural networks-based rolling bearing fault diagnosis. Microelectron. Reliab. **75**, 327–333 (2017). https://doi.org/10.1016/j.microrel.2017.03.006

95. Liang, G., Naipeng, L., Feng, J., Yaguo, L., Jing, L.: A recurrent neural network based health indicator for remaining useful life prediction of bearings. Neurocomputing **240**, 98–109 (2017)

96. Li, Z., Wu, D., Hu, C., Terpenny, J.: An ensemble learning-based prognostic approach with degradation-dependent weights for remaining useful life prediction. Reliab. Eng. Syst. Saf. **000**, 1–13 (2018). https://doi.org/10.1016/j.ress.2017.12.016

97. Choo, B.Y., Adams, S.C., Weiss, B.A., Marvel, J.A., Beling, P.A.: Adaptive multi-scale prognostics and health management for smart manufacturing systems. Int. J. Progn. Health Manag. **7**, 014 (2016)

98. Jia, F., Lei, Y., Guo, L., Lin, J., Xing, S.: A neural network constructed by deep learning technique and its application to intelligent fault diagnosis of machines. Neurocomputing **272**, 619–628 (2018). https://doi.org/10.1016/j.neucom.2017.07.032

 99. Shao, H., Jiang, H., Lin, Y., Li, X.: A novel method for intelligent fault diagnosis of rolling
     bearings using ensemble deep auto-encoders. Mech. Syst. Signal Process. **102**, 278–297
     (2018). https://doi.org/10.1016/j.ymssp.2017.09.026
100. Rajeswaran, N., Lakshmi Swarupa, M., Sanjeeva Rao, T., Chetaswi, K.: Hybrid artificial
     intelligence based fault diagnosis of SVPWM voltage source inverters for induction motor.
     Mater. Today Proc. **5**, 565–571 (2018). https://doi.org/10.1016/j.matpr.2017.11.119
101. Zhang, W., Li, C., Peng, G., Chen, Y., Zhang, Z.: A deep convolutional neural network
     with new training methods for bearing fault diagnosis under noisy environment and different
     working load. Mech. Syst. Signal Process. **100**, 439–453 (2018). https://doi.org/10.1016/j.
     ymssp.2017.06.022
102. Para, J., Del Ser, J., Nebro, A.J., Zurutuza, U., Herrera, F.: Analyze, Sense, Preprocess, Predict,
     Implement, and Deploy (ASPPID): an incremental methodology based on data analytics for
     cost-efficiently monitoring the industry 4.0. Eng. Appl. Artif. Intell. **82**, 30–43 (2019). https://
     doi.org/10.1016/J.ENGAPPAI.2019.03.022
103. Mourtzis, D., Vlachou, E.: A cloud-based cyber-physical system for adaptive shop-floor
     scheduling and condition-based maintenance. J. Manuf. Syst. **47**, 179–198 (2018). https://
     doi.org/10.1016/j.jmsy.2018.05.008
104. Paul Herrera: Decision Makers Must Prioritize Data Analytics In 2022. https://www.for
     bes.com/sites/forbesbusinesscouncil/2022/02/25/decision-makers-must-prioritize-data-ana
     lytics-in-2022/
105. Russom, P.: Big data analytics. TDWI Best Pract. Rep. Fourth Q. **6** (2011). https://doi.org/10.
     1109/ICCICT.2012.6398180
106. Vazan, P., Janikova, D., Tanuska, P., Kebisek, M., Cervenanska, Z.: Using data mining methods
     for manufacturing process control. IFAC-PapersOnLine. **50**, 6178–6183 (2017). https://doi.
     org/10.1016/j.ifacol.2017.08.986
107. Mourtzis, D., Vlachou, E., Milas, N.: Industrial big data as a result of IoT adoption in
     manufacturing. Procedia CIRP. **55**, 290–295 (2016). https://doi.org/10.1016/j.procir.2016.
     07.038
108. Matsas, M., Pintzos, G., Kapnia, A., Mourtzis, D.: An integrated collaborative platform for
     managing product-service across their life cycle. Procedia CIRP. **59**, 220–226 (2017). https://
     doi.org/10.1016/J.PROCIR.2016.09.009
109. Deng, S., Yeh, T.H.: Using least squares support vector machines for the airframe structures
     manufacturing cost estimation. Int. J. Prod. Econ. **131**, 701–708 (2011). https://doi.org/10.
     1016/j.ijpe.2011.02.019
110. Yeh, T.H., Deng, S.: Application of machine learning methods to cost estimation of product
     life cycle. Int. J. Comput. Integr. Manuf. **25**, 340–352 (2012). https://doi.org/10.1080/095
     1192X.2011.645381
111. Kassner, L., Gröger, C., Mitschang, B., Westkämper, E.: Product life cycle analytics—next
     generation data analytics on structured and unstructured data. Procedia CIRP. **33**, 35–40
     (2015). https://doi.org/10.1016/J.PROCIR.2015.06.008
112. Kumar, A., Shankar, R., Choudhary, A., Thakur, L.S.: A big data MapReduce framework for
     fault diagnosis in cloud-based manufacturing. Int. J. Prod. Res. **54**, 7060–7073 (2016). https://
     doi.org/10.1080/00207543.2016.1153166
113. Osemwegie, O., Okokpujie, K., Nkordeh, N., Ndujiuba, C., John, S., Stanley, U.: Performance
     benchmarking of key-value store NoSQL databases. Int. J. Electr. Comput. Eng. (IJECE) **8**,
     5333–5341 (2018). https://doi.org/10.11591/ijece.v8i6.pp533-5431
114. Chebotko, A., Kashlev, A., Lu, S.: A big data modeling methodology for Apache Cassandra.
     In: Proceedings—2015 IEEE International Congress on Big Data, BigData Congress 2015,
     pp. 238–245 (2015). https://doi.org/10.1109/BIGDATACONGRESS.2015.41
115. Chadwick, S.: 2015 capital investment in the market research and analytics sector. Res. World.
     2016, 9–13 (2016). https://doi.org/10.1002/RWM3.20318
116. Nakagawa, E.Y., Antonino, P.O., Schnicke, F., Capilla, R., Kuhn, T., Liggesmeyer, P.: Industry
     4.0 reference architectures: state of the art and future trends. Comput. Ind. Eng. **156**, 107241
     (2021). https://doi.org/10.1016/j.cie.2021.107241

117. Galati, F., Bigliardi, B.: Industry 4.0: Emerging themes and future research avenues using a text mining approach. Comput. Ind. **109**, 100–113 (2019). https://doi.org/10.1016/J.COMPIND.2019.04.018
118. Givehchi, O., Landsdorf, K., Simoens, P., Colombo, A.W.: Interoperability for industrial cyber-physical systems: an approach for legacy systems. IEEE Trans. Ind. Inform. **13**, 3370–3378 (2017). https://doi.org/10.1109/TII.2017.2740434
119. Bicaku, A., Schmittner, C., Delsing, J., Maksuti, S., Palkovits-Rauter, S., Tauber, M., Matischek, R., Mantas, G., Thron, M.: Towards Trustworthy End-to-End Communication in Industry 4.0 SECCRIT-SEcure Cloud computing for CRitical infrastructure IT View project Towards Trustworthy End-to-End Communication in Industry 4.0. (2017)
120. Fraile, F., Tagawa, T., Poler, R., Ortiz, A.: Trustworthy industrial IoT gateways for interoperability platforms and ecosystems. IEEE Internet Things J. **5**, 4506–4514 (2018). https://doi.org/10.1109/JIOT.2018.2832041
121. Petroulakis, N.E., Lakka, E., Sakic, E., Kulkarni, V., Fysarakis, K., Somarakis, I., Serra, J., Sanabria-Russo, L., Pau, D., Falchetto, M., Presenza, D., Marktscheffel, T., Ramantas, K., Mekikis, P.V., Ciechomski, L., Waledzik, K.: SEMIoTICS architectural framework: end-to-end security, connectivity and interoperability for industrial IoT. In: Global IoT Summit, GIoTS 2019—Proceedings (2019). https://doi.org/10.1109/GIOTS.2019.8766399
122. AUTOSAR: AUTomotive Open System ARchitecture (AUTOSAR). https://www.autosar.org/
123. U.S. Department of Transportation: Architecture Reference for Cooperative and Intelligent Transportation (ARC-IT). https://www.arc-it.net/html/architecture/architecture.html
124. Open Group: SOA Reference Architecture. http://www.opengroup.org/soa/source-book/soa_refarch/
125. O'Donovan, P., Leahy, K., Bruton, K., O'Sullivan, D.T.J.: Big data in manufacturing: a systematic mapping study. J Big Data. **2**, 1–22 (2015). https://doi.org/10.1186/S40537-015-0028-X/TABLES/8
126. Joint Research Centre: AI Watch. Defining Artificial Intelligence 2.0. (2021)
127. Patel, P., Ali, M.I., Sheth, A.: From raw data to smart manufacturing: AI and semantic web of things for industry 4.0. IEEE Intell. Syst. **33**, 79–86 (2018). https://doi.org/10.1109/MIS.2018.043741325
128. Khosravani, M.R., Nasiri, S., Weinberg, K.: Application of case-based reasoning in a fault detection system on production of drippers. Appl. Soft Comput. **75**, 227–232 (2019). https://doi.org/10.1016/J.ASOC.2018.11.017
129. Costa Silva, G., Carvalho, E.E.O., Caminhas, W.M.: An artificial immune systems approach to Case-based Reasoning applied to fault detection and diagnosis. Expert. Syst. Appl. **140**, 112906 (2020). https://doi.org/10.1016/J.ESWA.2019.112906
130. Alexopoulos, K., Nikolakis, N., Chryssolouris, G.: Digital twin-driven supervised machine learning for the development of artificial intelligence applications in manufacturing **33**, 429–439 (2020). https://doi.org/10.1080/0951192X.2020.1747642
131. Alexopoulos, K., Sipsas, K., Xanthakis, E., Makris, S., Mourtzis, D.: An industrial Internet of things based platform for context-aware information services in manufacturing. Int. J. Comput. Integr. Manuf. **31**, 1111–1123 (2018). https://doi.org/10.1080/0951192X.2018.1500716
132. Efthymiou, K., Sipsas, K., Mourtzis, D., Chryssolouris, G.: On knowledge reuse for manufacturing systems design and planning: A semantic technology approach. CIRP J. Manuf. Sci. Technol. **8**, 1–11 (2015). https://doi.org/10.1016/j.cirpj.2014.10.006
133. Wan, J., Yin, B., Li, D., Celesti, A., Tao, F., Hua, Q.: An ontology-based resource reconfiguration method for manufacturing cyber-physical systems. IEEE/ASME Trans. Mechatron. **23**, 2537–2546 (2018). https://doi.org/10.1109/TMECH.2018.2814784
134. Liu, H.C., You, J.X., Li, Z.W., Tian, G.: Fuzzy Petri nets for knowledge representation and reasoning: a literature review. Eng. Appl. Artif. Intell. **60**, 45–56 (2017). https://doi.org/10.1016/J.ENGAPPAI.2017.01.012
135. Kim, J.H., Wang, L.S., Putta, K., Haghighi, P., Shah, J.J., Edwards, P.: Knowledge based design advisory system for multi-material joining. J. Manuf. Syst. **52**, 253–263 (2019). https://doi.org/10.1016/J.JMSY.2019.03.003

136. Wang, W.M., Peng, X., Zhu, G.N., Hu, J., Peng, Y.H.: Dynamic representation of fuzzy knowledge based on fuzzy petri net and genetic-particle swarm optimization. Expert Syst. Appl. **41**, 1369–1376 (2014). https://doi.org/10.1016/J.ESWA.2013.08.034

137. Alexopoulos, K., Makris, S., Xanthakis, V., Sipsas, K., Chryssolouris, G.: A concept for context-aware computing in manufacturing: the white goods case. Int. J. Comput. Integr. Manuf. **29**, 839–849 (2016). https://doi.org/10.1080/0951192X.2015.1130257

138. Belkadi, F., Dhuieb, M.A., Aguado, J.V., Laroche, F., Bernard, A., Chinesta, F.: Intelligent assistant system as a context-aware decision-making support for the workers of the future. Comput. Ind. Eng. **139**, 105732 (2020). https://doi.org/10.1016/J.CIE.2019.02.046

139. Pizoń, J., Lipski, J.: Manufacturing process support using artificial intelligence. Appl. Mech. Mater. **791**, 89–95 (2015). https://doi.org/10.4028/www.scientific.net/amm.791.89

140. Makris, S., Pintzos, G., Rentzos, L., Chryssolouris, G.: Assembly support using AR technology based on automatic sequence generation. CIRP Ann. Manuf. Technol. **62**, 9–12 (2013). https://doi.org/10.1016/j.cirp.2013.03.095

141. Rentzos, L., Papanastasiou, S., Papakostas, N., Chryssolouris, G.: Augmented reality for human-based assembly: using product and process semantics. In: IFAC Proceedings Volumes (IFAC-PapersOnline), pp. 98–101. Elsevier (2013)

142. Zhang, J., Ong, S.K., Nee, A.Y.C.: RFID-assisted assembly guidance system in an augmented reality environment. Int. J. Prod. Res. **49**, 3919–3938 (2011). https://doi.org/10.1080/002 07543.2010.492802

143. Ong, S.K., Wang, Z.B.: Augmented assembly technologies based on 3D bare-hand interaction. CIRP Ann. Manuf. Technol. **60**, 1–4 (2011). https://doi.org/10.1016/j.cirp.2011.03.001

144. Sääski, J., Salonen, T., Hakkarainen, M., Siltanen, S., Woodward, C., Lempiäinen, J.: Integration of design and assembly using augmented reality. In: IFIP International Federation for Information Processing, pp. 395–404. Springer, Boston, MA (2008)

145. Pentenrieder, K., Bade, C., Doil, F., Meier, P.: Augmented reality-based factory planning—an application tailored to industrial needs. In: 2007 6th IEEE and ACM International Symposium on Mixed and Augmented Reality, ISMAR. pp. 31–42. IEEE Computer Society (2007)

146. Makris, S., Karagiannis, P., Koukas, S., Matthaiakis, A.S.: Augmented reality system for operator support in human–robot collaborative assembly. CIRP Ann. Manuf. Technol. **65**, 61–64 (2016). https://doi.org/10.1016/j.cirp.2016.04.038

147. Monostori, L.: AI and machine learning techniques for managing complexity, changes and uncertainties in manufacturing. In: Engineering Applications of Artificial Intelligence, pp. 277–291. Pergamon (2003)

148. Dash, R., Mcmurtrey, M., Rebman, C., Kar, U.K.: Application of artificial intelligence in automation of supply chain management. J. Strateg. Innov. Sustain. **14**, 43 (2019). https://doi.org/10.33423/jsis.v14i3.2105

149. Kumara, S.R.T., Ranjan, P., Surana, A., Narayanan, V.: Decision making in logistics: a chaos theory based analysis. CIRP Ann. Manuf. Technol. **52**, 381–384 (2003). https://doi.org/10.1016/S0007-8506(07)60606-4

150. Louly, M.A., Dolgui, A., Hnaien, F.: Supply planning for single-level assembly system with stochastic component delivery times and service-level constraint. Int. J. Prod. Econ. **115**, 236–247 (2008). https://doi.org/10.1016/j.ijpe.2008.06.005

151. Ilie-Zudor, E., Monostori, L.: Agent-based framework for pre-contractual evaluation of participants in project-delivery supply-chains. Assem. Autom. **29**, 137–153 (2009). https://doi.org/10.1108/01445150910945598

152. Saint Germain, B., Valckenaers, P., Verstraete, P., Hadeli, Van Brussel, H.: A multi-agent supply network control framework. Control Eng. Pract. **15**, 1394–1402 (2007). https://doi.org/10.1016/j.conengprac.2006.12.003

153. Kara, A., Dogan, I.: Reinforcement learning approaches for specifying ordering policies of perishable inventory systems. Expert Syst. Appl. **91**, 150–158 (2018). https://doi.org/10.1016/J.ESWA.2017.08.046

154. Mourtzis, D., Papakostas, N., Makris, S., Xanthakis, V., Chryssolouris, G.: Supply chain modeling and control for producing highly customized products. CIRP Ann. Manuf. Technol. **57**, 451–454 (2008). https://doi.org/10.1016/j.cirp.2008.03.106

155. Chryssolouris, G., Makris, S., Xanthakis, V., Mourtzis, D.: Towards the Internet-based supply chain management for the ship repair industry. Int. J. Comput. Integr. Manuf. **17**, 45–57 (2004). https://doi.org/10.1080/0951192031000080885

156. Mourtzis, D., Doukas, M., Psarommatis, F.: Design and operation of manufacturing networks for mass customisation. CIRP Ann. Manuf. Technol. **62**, 467–470 (2013). https://doi.org/10.1016/j.cirp.2013.03.126

157. Yu, T.L., Yassine, A.A., Goldberg, D.E.: An information theoretic method for developing modular architectures using genetic algorithms. Res. Eng. Des. **18**, 91–109 (2007). https://doi.org/10.1007/s00163-007-0030-1

158. Takai, S., Yang, T., Cafeo, J.A.: A Bayesian method for predicting future customer need distributions. Concurr. Eng. **19**, 255–264 (2011). https://doi.org/10.1177/1063293X11418135

159. Liu, Z., Guo, S., Wang, L., Du, B., Pang, S.: A multi-objective service composition recommendation method for individualized customer: hybrid MPA-GSO-DNN model. Comput. Ind. Eng. **128**, 122–134 (2019). https://doi.org/10.1016/J.CIE.2018.12.042

160. Makris, S., Zoupas, P., Chryssolouris, G.: Supply chain control logic for enabling adaptability under uncertainty. Int. J. Prod. Res. **49**, 121–137 (2011). https://doi.org/10.1080/00207543.2010.508940

161. Pandremenos, J., Chryssolouris, G.: A neural network approach for the development of modular product architectures. Int. J. Comput. Integr. Manuf. **24**, 879–887 (2011). https://doi.org/10.1080/0951192X.2011.602361

162. Chen, T.: Estimating unit cost using agent-based fuzzy collaborative intelligence approach with entropy-consensus. Appl. Soft Comput. **73**, 884–897 (2018). https://doi.org/10.1016/J.ASOC.2018.09.036

163. Chou, J.S., Cheng, M.Y., Wu, Y.W., Tai, Y.: Predicting high-tech equipment fabrication cost with a novel evolutionary SVM inference model. Expert Syst. Appl. **38**, 8571–8579 (2011). https://doi.org/10.1016/J.ESWA.2011.01.060

Printed in the United States
by Baker & Taylor Publisher Services